T0290939

Praise for *What Walks This Way*

"A collection of beautifully written stories about the author's quest to deepen her connection with nature."

—Jonah Evans, CyberTracker evaluator, nongame and rare species program director for Texas Parks and Wildlife

"Reading *What Walks This Way* is like going on a slow and restorative hike with a close friend. The book is filled with both personal stories and insight into the natural history of North American mammals, giving the reader a guide to more deeply witness and love the lives of wildlife around us. Sharman Apt Russell offers an antidote to defaunation, one dusty footprint at a time."

—Emily Burns, program director, Sky Island Alliance

"An enjoyable ramble through the world of citizen trackers. I found a quiet pleasure in reading Russell's book about lay naturalists who are developing simple, nontechnological skills far beyond those of most of our institutional biologists."

—Harley Shaw, author of *Soul Among Lions: The Cougar as Peaceful Adversary*

"I love the montage used to build human-sign-animal connections in each chapter—the vignettes of straight-up tracking info, social-psychological dynamics, history, and stripped-down conversations. It's great stuff."

—David Mattson, founder, *Grizzly Times*

"Take this book with you and go outside. Bend down and look closely. Let Sharman Apt Russell instruct you in how to remember your place in the family of beings we share this planet with. I can't think of a better mentor than this seasoned writer who conveys the gravity of our and their situation but at the same time injects heart and humor into this beautiful field guide to our animal kin."

—Lorraine Anderson, editor of *Sisters of the Earth: Women's Prose and Poetry About Nature*

"After reading this book, I will never again walk along a sandy wash and see it the same way. Sharman Apt Russell takes the reader on a poignant yet playful journey, following not only tracks but also the animals' sagas and status as they prowl and tromp through our modern world. Her compassion for their fate is equaled by the humor and wonder that she finds with each step along the way."

—Anne Lane Hedlund, author of *Navajo Weaving in the Twentieth Century* and coauthor of *Navajo Weavers of the American Southwest*

"Not a field guide, not a textbook, but a series of stories of introduction and observation, of facts and figures interspersed with story and connection. Sharman Apt Russell sets the stage, complete with backstory and current status, of more than two dozen wild critters one might encounter in North America. With support from Kim A. Cabrera's collection of track photos, this book weaves through her unique experience connecting with the natural world. A thoroughly enjoyable book!"

—Shane Hawkins, Tracker Certification North America and Original Wisdom

"I live on the edge of a desert city on land traversed by coyotes, deer, javelinas, bobcats, all manner of snakes, the occasional mountain lion and black bear, a dozen species of rodents and another dozen species of lizards, and, overhead, a hundred kinds of birds. As Sharman Apt Russell writes in this luminous portrait, they 'do not want us seeing into their secret lives.' But see and learn we must if we are to help them survive. This peerless book is just the place to start."

—Gregory McNamee, author of *Gila: The Life and Death of an American River*

"Sharman Apt Russell loves tracking, observing what walked this way. She includes requisite graphics and detailed discussions to give enthusiasts a good chance to succeed at identifying mammalian spoor. Chapters run the gamut from big ones (cougar) to small ones (rodents). Beyond identification essentials, Russell also lays out interesting context and background on species. All in all, this book shines a light, figuratively and literally, on the generally ignored, unseen evidence of animals that, when observed, fires the imagination."

—Jim Furnish, author of *Toward a Natural Forest* and retired deputy chief, U.S. Forest Service

"An imaginative and personal track-and-sign conversation ripe with insight and wisdom."

—David Moskowitz, Tracker Certification North America, biologist, photographer, and outdoor educator

"I have been following Sharman Apt Russell's work for decades and always end up seeing the world differently as a result of the tracks she leaves across the mesas, along the riverbanks, and in the sand of southwest New Mexico. Few writers match her ability to combine information with passion and perception. She is a master storyteller, and *What Walks This Way* is an astounding book."

—M. John Fayhee, author of *A Long Tangent* and *Smoke Signals*

What Walks This Way

WHAT WALKS THIS WAY

Discovering the Wildlife Around Us Through Their Tracks and Signs

SHARMAN APT RUSSELL

ILLUSTRATIONS AND PHOTOS BY KIM A. CABRERA

Columbia University Press
New York

Columbia University Press
Publishers Since 1893
New York   Chichester, West Sussex
cup.columbia.edu

Library of Congress Cataloging-in-Publication Data
Names: Russell, Sharman Apt, author. | Cabrera, Kim, illustrator, photographer.
Title: What walks this way : discovering the wildlife around us through their tracks and signs / Sharman Apt Russell ; illustrations and photos by Kim Cabrera.
Description: New York : Columbia University Press, [2024] | Includes bibliographical references and index. Identifiers: LCCN 2024003379 | ISBN 9780231215985 (hardback) | ISBN 9780231215992 (trade paperback) | ISBN 9780231561075 (ebook)
Subjects: LCSH: Animal tracks. | Animal droppings. | Habitat (Ecology) | Habitat conservation.
Classification: LCC QL768 .R85 2024 | DDC 591.47/9—dc23/eng/20240405
LC record available at https://lccn.loc.gov/2024003379

Cover design: Henry Sene Yee

To Sonnie
—Sharman

To my mom, Karen Cabrera, and all the cats over the years
—Kim

Contents

What Walks This Way

1

Ribbon of Life

For almost a decade I looked at wildlife tracks with my friend Sonnie, at first casually and then seriously, as we helped monitor black bears and mountain lions in New Mexico for a southwestern environmental group. Later we studied together in an evaluation process that certifies wildlife trackers in the field. Sonnie was diabetic, with lung damage from years of smoking, and the pace of identifying track and sign suited her well. A slow walk, stop, bend, look. Four dainty toes, suggestion of fur, and that wavy line at the bottom of the palm pad. A gray fox passed this way. Over here: four teardrop toes, one middle toe extended past the other, palm pad more than two inches long. Mountain lion, possibly male.

This weird squiggle? A darkling beetle.

A sinuous curve with smudges on either side? Some kind of lizard.

One spring day Sonnie turned to another friend with whom she was hiking, or likely strolling, musing, bird watching, and said, "It's so beautiful here," before she fell to the ground. She was gone in that instant. Her memorial, long delayed because of COVID-19, featured a ravishing blue sky of cumulus clouds and a red-tailed hawk. I quoted from Aldo Leopold, a seminal environmentalist she had often quoted, too: "There are some who can live without wild things and some who cannot."[1]

Sonnie knew which one she was. So do most of my friends. The pleasure we get in living with wild things feels pure and uncomplicated. *A gray fox passed this way.* In learning how to identify track and sign, we enjoy that democratic thrill, something almost anyone can have almost anywhere, something you can take up at almost any age or physical condition. We bend down to look, using our imagination, our mirror neurons, our reading glasses, matching marks and shapes to meaning and story. As hunters and gatherers, we depended on this skill. Perhaps that's why books and emails are so familiar.

Our pleasure, obviously, is not reciprocated. The wild animals who leave their tracks do not care about us, except for the desire to be invisible. They want to wind secretly past my house while I am sleeping, past your house while you are sleeping. They want to be unseen, unnoticed, unloved.

A glorious exception is scat. More on that later.

The wild animals on this Earth do not want us seeing into their secret lives, and we know why. We've killed so many of them already. It's not just extinction. It's the loss of abundance. In the past fifty years, the populations of more than five thousand species of fish, reptiles, amphibians, birds, and mammals have dropped by almost 70 percent. Humans and our domesticated livestock now account for 96 percent of the weight of mammals worldwide. Wild mammals add up to 4 percent. The scientific term is *defaunation.* At first that word feels awkward, stilted, like marbles in the mouth. Defaunation. Then it becomes exactly right.[2]

In our years of tracking, Sonnie and I began to get a better sense of the animals who live where we live. Some of this information was for that southwestern environmental group that wanted to know about not only black bears and mountain lions but also any threatened and endangered species, such as ornate box turtles and New Mexico meadow jumping mice. In particular, Sonnie and I dreamed of finding the track of the endangered and reintroduced Mexican gray wolf, and I remember literal dreams, waking up to think this had just happened. I had been in a red rock canyon with high yellow cliffs. No, I had been running along a trail, watching the ground and then stopping, stopped as if someone held my arm, before that symmetrical shape and those sharp claw marks. A meadow jumping mouse would have been as exciting, the splayed toes so endearingly small. From one season to the next, we noticed when striped skunks and gray foxes suddenly declined, likely because of disease. We were surprised when rabbits and hares seemed to decline, too.

I began to wonder about the bigger picture. New Mexico is my home. The Southwest is my home. North America is my home. I realized I knew surprisingly little about the numbers and status of so many animals resonant in my imagination and psyche. Are badgers okay? Do we have enough pika? Or caribou? What does defaunation look like in northern North America—the United States and Canada—and how will global warming affect the future?

One intention of this book is to answer that question.

Which brings up another question: How do we know? Who is keeping track, forgive the pun, of our abundance of wildlife? The obvious answer would be elected governments and their bureaucracies, as well as conservation and environmental groups. But it's not so simple—first because wildlife is actually pretty hard to monitor and manage, something we might have predicted as inherent in those two words, *wild* and *life*, but also because people have differing ideas about wildlife, defining and valuing it from different perspectives.

One perspective is long-standing. The Europeans who colonized North America saw wildlife mostly as a resource, something to be consumed, or a threat, something to be destroyed. Later, as populations of game animals crashed because of habitat loss and overhunting, wildlife also became something to protect. Today, in both the United States and Canada, governments hold wildlife as a "public trust asset" for the benefit of present and future generations, an idea rooted in the forest laws of Great Britain. Federal, state, provincial, and territorial agencies regulate species labeled as protected furbearers and game animals, as well as threatened and endangered species and migratory species protected under treaties with other countries. Everywhere in North America, there are rules about which wild animals can be killed at certain times and in certain ways, which can be killed at any time, and which cannot be harmed at all.

In terms of keeping track, we try to quantify the species we are losing. In 2023, perhaps 350 black-footed ferrets, 250 Vancouver Island marmots, and 240 Mexican gray wolves remained in the wild. Arguably, we are most interested in the species we hunt. That same year, the state of New Hampshire estimated it had 3,300 moose. New York believed it had between 6,000 and 8,000 black bears. Wyoming had about 380,000 pronghorn, a significant reduction from previous years because of severe winter weather.

Another vision of wildlife is one I have been exploring most of my life. This vision does not exclude hunting, fishing, or trapping. This vision also supports protection and management. But the point of all that is community, not consumption. What we desire is connection, not control.

Indigenous people in North America have been saying this forever. More recently, environmentalists and scientists are saying the same thing. Many people, no matter how they self-identify, feel the need of this kind of relationship. Certainly they are in community with their cat or dog—but also with the opossum in the backyard, or that white-tailed deer browsing the willow. These animals are individual beings, conscious and thrumming with specific existence, mysterious because they are not human beings, and a little glamorous for the same reason.

Our community of wild animals is huge and just outside the door. Some of the best days Sonnie and I had identifying track and sign were when we turned from mammals to insects, reptiles, and birds. Often enough, we first saw the caterpillar trundle across sand, leaving behind the distinct tread that matched up with our *Tracks and Sign of Insects and Other Invertebrates: A Guide to North American Species.* A bullfrog hopped into the river, and there along the muddy bank was the imprint of five angled toes and a plump bottom resting between two front feet. An annoyed blue heron had just lifted off moments before. Life, wild, everywhere.

For this book, however, I have turned back to mammals. That's partly because so many of them are otherwise unseen, winding past our houses while we are sleeping, deliberately and successfully undercover. I have included most of the common mammals in North America, with the majority of those ranging over large areas of the United States and Canada. Their abundance has been diminished, and that loss is felt deeply by many people. But loss is only half the story and not where I like to spend too much of my time. This, then, is a celebration of the wild animals that are still here—mountain lions and bears, foxes and weasels, badgers and wolves, creatures alive in our myth and lore and also in our forests and fields. This is equally and more so about skunks, raccoons, mice, rabbits, squirrels, and deer, those weedy species that reproduce quickly and have adapted to living beside us.

If these animals are familiar, if they sometimes seem curious and friendly, none of them are our domestics. They are the wild 4 percent that remain. Although I accept that they do not want to engage with me, I want to engage with them. I do not want to intrude on their secret lives; this is not a book about trailing or following animals. But I want to name them, ground squirrel or coyote or muskrat. I want to feel their presence. Because, after all, they were present just an hour or two ago—or yesterday or last week. *A gray fox passed this way.*

Or maybe a red fox, a species distantly related to the gray fox and more comfortable in our cities and suburbs. Gray foxes and red foxes coexist across the United States and southern Canada but don't interbreed. The track of a red fox would be larger than that of a gray fox, with more imprint of fur and a different shape of pad. For this particular blurred track, you'll need to get closer.

One of the first things you will do as a tracker is make a list of the animals in your area. All tracking experiences are local. Then, rather quickly, you'll begin to see these animals in their track and sign. Eventually you'll have this community, your neighbors, not so much memorized as cellular. They will become part of who you are, like the parts of your body: arms, legs, hands, feet. Or the furniture in your house, if you don't change your furniture very much. Or your favorite foods or songs you sang as a child.

You don't have to become an expert. But by the last chapter, I'd like you to see a print in the dirt and think feline or canine and then bobcat or domestic cat, fox or coyote. I want you to think raccoon. River otter. You might be wrong. You'll be savvy enough to say, as Sonnie and I often did, "I'm not really sure." You'll need to look at the pictures to do even that. You might try osmosis, something I have tried with so many field guides on my bookshelf. It hasn't worked yet, although I remain hopeful.

This, of course, is not a field guide. This is what you get before a field guide. My friend Sonnie would call this a love letter. I know. I know. I know. I know. The word *love* is much overused. Love, that psychic substratum, the measure of every good day's end. Did I see today the one I love? Did I do today what I love to do?

For myself, I am always trying to love the world more, slow down, bend down, look, and really see. I am not very good at this. Perhaps you share that feeling of incompetence, a kind of learning disability in the natural world. Perhaps this is a modern problem, or a cultural one. In any case, identifying the track of a spotted skunk helps. It's a cure for all kinds of inadequacy or despair. That may be an overstatement. But I don't think so.

2

Turn Your Dog's Paw Upside Down

I live in southwestern New Mexico, just a few miles from the three-million-acre Gila National Forest and Gila Wilderness, America's first wilderness, designated in 1924. The animals who might call this public land their home don't consider anything around here not their home. They travel purposefully over hills and into canyons, down the Gila River, up the Gila River, across fields, through my yard, onto my porch, and back into the trees along the irrigation ditch. Many of the tracks I find are only a few steps or a short walk away.

The humans who live on my country road are that odd mix typical of rural America. Since the 1960s, people who do not fit into mainstream society have been moving here, some staying, some not. A handful of second- and third-generation ranchers remain in this agricultural valley, although most of the water rights and land are now owned by an international mining company. As well as ranchers, a schoolteacher lives on my road, a freelance copy editor, an environmentalist with the Nature Conservancy, a retired postmaster, a retired city planner, a plumber, a writer. Perhaps half of us enjoy the skunk on the porch and the coyote in the grama grass. The other half shoot them.

Every day Michelle walks her three German shepherds down this road. Michelle is a tall striking woman with an abundance of dark hair who looks younger than she is and who carries in her straight back and muscular shoulders a certain force and energy. I don't think she is less flawed than the rest of us but just appears so. That impression is heightened by the three large dogs who walk, unleashed, by her side. Michelle is the mistress of these animals. She says "heel" and they stay close. She says "free" and they scatter over the undulations of prickly pear and mesquite, sniffing out holes, hoping for a cottontail or, even better, a jackrabbit.

And in this joyful world we live in, SHAZAM, sometimes there is a jackrabbit. A ripple in the space-time continuum. A flash of white. And the dogs are off!

Until Michelle calls them back. What happens next is something you don't see every day: the dogs do come back, almost immediately, even though they would rather not. If you are me, you stare in wonder. Americans own more than eighty million dogs, and the vast majority of them would keep chasing the jackrabbit.

Michelle's dogs come back because she has trained them and keeps training them. She is their family and their context. Michelle is a private person and wouldn't want me to write too much about her, but she does have a business in which she helps other people train their dogs, too, for the benefit of everyone. This training is about confidence and consistency as a human being, about trusting and being trusted as a dog. "You have to be the best person you can be," Michelle says.

The truth is that when I go on a walk or a run and see animal tracks, often they are domestic dog tracks—usually accompanied by a human footprint. This will be true even along a remote stretch of the Gila River, even if I am hiking miles away in the middle of the Gila Wilderness. Domestic dog tracks are distinct, and the tracks of Michelle's German shepherds are strikingly handsome: large, symmetrical, with deep, powerful-looking claw marks.

Turn the paw of a German shepherd upside down. The front and hind feet each have four fleshy toe pads that absorb the impact of running and leaping. A reduced fifth toe pad and claw are higher up on the front legs only. Under the toe pads are what trackers call the palm pad. Toe pads, more simply called toes, and palm pads—their presence, position, size, and shape—are major descriptors of a track. On a dog's track, the nails or claws often show as marks above the toes. Cats can retract their claws. Dogs cannot.

The inability to retract claws has had some impact on canine evolution.

Members of the order Carnivora appeared in North America about sixty million years ago. Those early carnivorans diverged into caniforms,

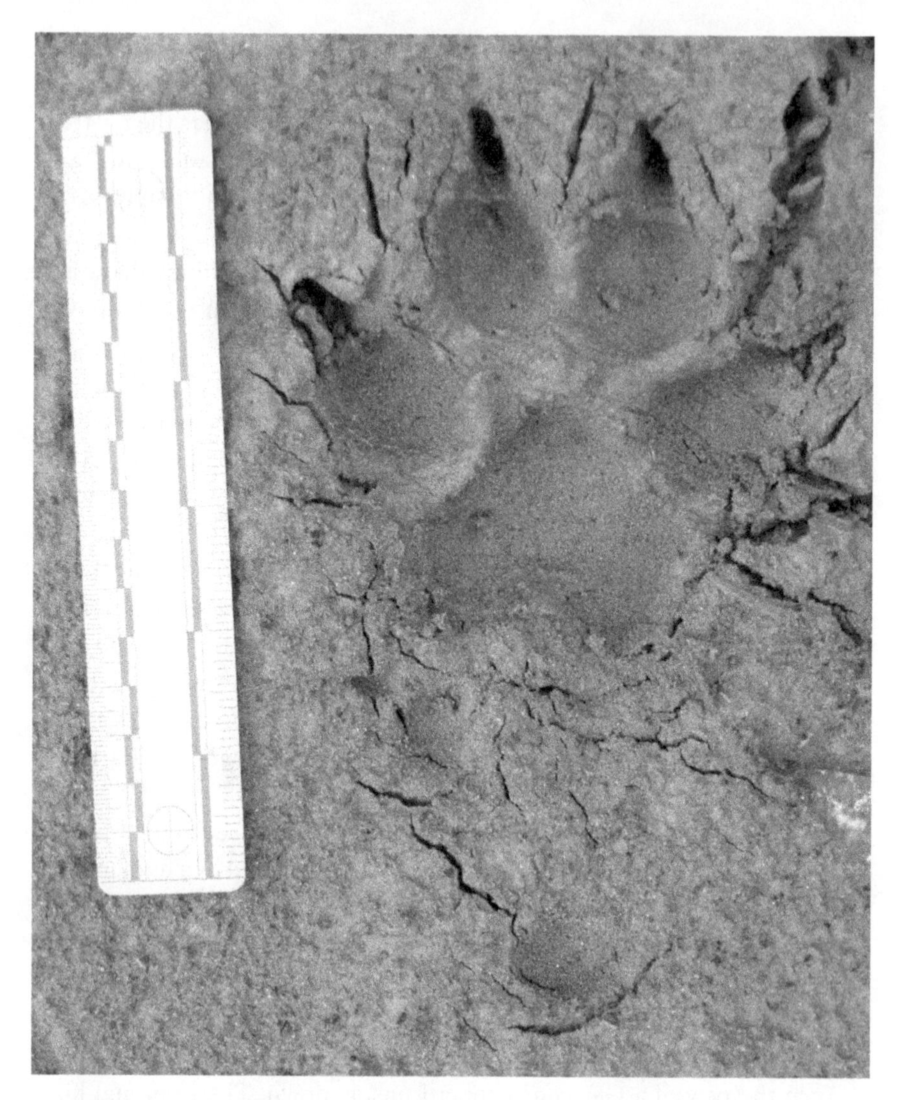

FIGURE 2.1 Domestic dog, right front

or doglike animals, and feliforms, or catlike animals. In particular, the Canidae, or dog family, seems to have originated in North America. Eventually, they encompassed dozens of species, all using their claws to keep traction on the ground as they hunted and ran down their prey.

Meanwhile, the feliforms also evolved, died out, moved around, and left big gaps in the fossil record. We first find the cat family, Felidae, which includes our modern felines, in Europe about twenty-five million years ago. Sometime after, those primitive cats began crossing the Bering land bridge into North America. For these early cats, retractable claws meant more silent and stealthier stalking. Retractable claws reduced the wear and tear on claws, meaning sharper claws for climbing trees and grabbing and holding onto other animals. In the next fifteen million years of competition between cats and dogs in North America, up to forty dog species went extinct.[1] Cat species were not affected—pretty much winning that contest.

The term *Bering land bridge* is misleading. We think of something relatively small, a walkway for pedestrians or a highway for cars. But the Bering land bridge connected present-day Russia, Canada, and Alaska. When the seas retreated during various ice ages—the water locked up by glaciers—this area could extend 620,000 square miles. For thousands of years at a time, these grasslands and forests supported grazing animals like mammoths, antelope, and horses. Horses, in particular, were wonderful prey for big cats and packs of dogs.

About five million years ago, members of Canidae left North America and began crossing the Bering land bridge to colonize Eurasia, then Europe, then Africa. Around the world, they evolved to become our current thirty-seven living species of canines, which include the jackal, the dhole, and the African wild dog. In this great radiation of the dog family, some animals that had evolved into new forms in Eurasia and Europe would later migrate back to North America. Some three million years ago, with the appearance of the Panama isthmus, canines in North America could travel south, where they diversified into existing species like the bush dog and the crab-eating fox.

How best to imagine this: leaving your home because you are compelled to leave home whenever there is a path to follow, a land bridge to cross or an island to swim toward, and then changing, morphing, becoming something new, and in the next blink of time, returning home because you are compelled to do that, too, whenever another path opens. And all around you, other species are compelled by the same instinct, moving and changing up and down the planet, which is changing as well, getting colder, hotter, even the continents still breaking apart a little.

What's remarkable, perhaps, are the canine species that didn't successfully migrate. The coyote evolved in North America and stayed. The gray fox evolved in North America and stayed. The kit fox and swift fox may have done the same.

Red foxes, on the other hand, began in North America, went to Eurasia and Europe, kept evolving, and came back across the Bering land bridge, moving southward into what is now the contiguous United States about 130,000 to 100,000 years ago. Today our American red fox is considered a species distinct from the European red fox. (Introduced into North America in the eighteenth century and and again in the early twentieth century, European red foxes died out and their genes are not significantly present in American red foxes.)[2] Arctic foxes also probably entered North America during the most recent ice age, roughly 100,000 to 10,000 years ago.

The gray wolf followed a similar pattern, going to Eurasia as some kind of canid, becoming a gray wolf perhaps in the Beringia region, spreading across the world, and returning to midcontinent North America about 100,000 years ago. Other wolves may have been here, too, like the red wolf—although there is a strong argument that red wolves are a subspecies of the gray wolf or the gray wolf hybridized with coyotes. In truth, most wolves, even those living in Italy or India (although not Africa) are a subspecies of the gray wolf. This animal traveled a lot and has one of the widest distributions of any mammal species.

In North America, red foxes and gray wolves shared the landscape with pack-hunting dholes and dire wolves. They coexisted with giant sloths, giant beavers, giant bison, camels, large-headed llamas, American lions, scimitar cats, cheetahs, cave bears, giant bears, saber-toothed cats, horses, tapirs, stag-moose, gomphotheres, mastodons, and mammoths. There was so much to see, so many smells and sounds. The land rippled with life.

By the end of the Pleistocene, some 12,000 calendar years ago, thirty-eight genera and more than fifty species in North America were gone—every species listed above, and many more. Six genera—dholes, horses, spectacled bears, capybara, tapirs, and saigas—still lived elsewhere. Scientists have long debated the reasons behind these losses, most of which seemed to come quickly in the space of a thousand years. Some researchers question this timeline, pointing to the possibility of earlier or later extinctions ranging over a longer time period. Some emphasize a warming climate. The prevalent theory points to humans, who also crossed the land bridge or sailed the North American coastline in boats. These humans spread over the continent, killing large animals that had never known the reach of a spear. Perhaps disease played a role. Perhaps there were multiple reasons.[3]

The hunters who came to this species-rich landscape brought one new species of their own: the domesticated dog. The domestication of dogs probably began somewhere in Eurasia with a few wolves, born a little friendlier than other wolves, hanging around campsites and begging for food. Normally we don't care for wolves. But we do respond to friendliness, yearning looks, and puppy eyes. The domesticated or domesticating dogs learned our language, and we learned theirs. The benefit was mutual. Dogs helped us hunt, travel across snow, and defend ourselves. In an emergency, they could be eaten.

Hunters and their dogs probably came to North America more than 20,000 years ago. Exactly when is another scientific debate. Domestic dogs were introduced again in the fifteenth century, when Europeans arrived with breeds that would largely replace those of the Native Americans. This replacement was the result of new infectious canine diseases, as well as the persecution of native dogs.

The eleven species of canines in North America—domestic dog, coyote, gray fox, red fox, arctic fox, swift fox, kit fox, island fox, gray wolf, red wolf, eastern wolf—are not rigid categories. Isolated in the Channel Islands off southern California, the island fox is clearly descended from the gray fox. The swift fox and the kit fox may be the same species. The red wolf and the eastern wolf may be an ancestral mix of gray wolf and coyote. The now larger coyote in the eastern part of the United States and Canada is definitively a mix of coyote, wolf, and dog—animals perfectly capable of breeding with each other and producing fertile offspring. The concept of what defines a species is not rigid either.

All these canine tracks share some common features.

They usually register four toes.

They are symmetrical. One half of the track looks roughly like the other half.

They have a typical pattern in what trackers call the negative space. Like a rubber stamp, this is the area between the toe pads and palm pad when the foot touches a surface. In canines, you can usually draw an X in this negative space without bisecting any of the pads. *X for Xylophone* is my mnemonic. (Sometimes that negative space looks more like an H.)

Sometimes that negative space will show as a raised mound or pyramid.

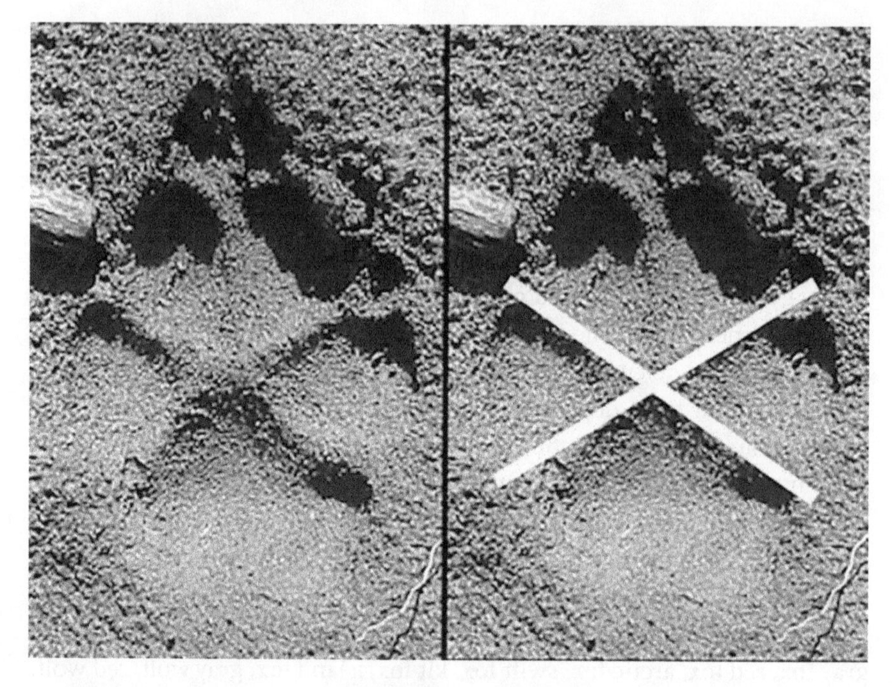

FIGURE 2.2 Hind track of golden retriever named Holly

Canine tracks also have palm pads that are relatively small compared to the size of the track. The top of the palm pad in a canine has one lobe and can look like a point. The bottom of the palm pad has two or three lobes.

These tracks usually, but not always, show the impression of claws or nails. A domestic dog's claws are blunter and heavier and typically more visible than those of wild canines like coyotes and foxes.

Domestic dogs, of course, come in many different breeds, and their tracks reflect that variety. Standard schnauzers and Tibetan mastiffs have "cat feet," round and compact, good for keeping a grip on slippery surfaces and efficient in terms of the energy required to lift each foot. Greyhounds and whippets have "hare feet," with two center toes longer than the other toes in an elongated paw, less efficient in terms of lifting a foot but providing more speed for sudden sprinting. The toes of a Labrador retriever and a Newfoundland are webbed. Norwegian lundehunds have six toes.

Along my country road, I see a variety of dogs all mixed up, mongrel mixes of pit bulls and hounds and heelers and retrievers, most running free and none so well-behaved as Michelle's German shepherds. I am not afraid of dogs, who almost always have that longing look, "Can't we be friends?" Even barking dogs usually transmit, "This is for show! I'm compelled to do this! Can't we be friends!"

The exception is when I'm bicycling. Turning, flashing wheels seem to trigger a greater protective urgency in dogs. They may remind dogs of rabbits or hares. The dogs give chase. They dart and nip. And then I *am* afraid, of falling as much as being bitten, slamming my knee into the ground, entangling with spokes and gears—a trip to the ER and the entire day wasted. Quickly, too, the knee swells and maybe the ankle. Years later, arthritis will seek out those damaged parts of the body. These thoughts are interrupted by the more immediate sensation of being attacked, the darting movement of a predator trying to make me panic, hoping to tire me out. Enough nips and maybe I'll die of shock or blood loss. Soon someone in the pack will jump on my nose, suffocating me. Someone in the pack will grab a leg, bringing me down. That's when they rip the intestines from your belly. By now I am screaming, "BAD BAD DOG!" I don't care if this is poor training technique, if I am not being the best person I can be.

Eighty percent of dog bites happen at home, not on a bicycle. The CDC estimates that more than 4.5 million people in America are bitten by a domestic dog each year, with many of those family pets biting a child. Most fatalities and severe injuries are also in young children. Almost all dog bites have to do with poor management of dogs by humans.

My tracking friend Sonnie had a rescue dog, Pumpkin, a lab mix. Pumpkin was not especially well trained, but she was sweet and friendly and very happy to come with us on our walks. We kept her on a leash at all times because otherwise we would never see her again. Sonnie and Pumpkin loved each other as much as I have ever seen a human and a dog in love. Sonnie would put her nose right down to Pumpkin's nose. They looked into each other's eyes. Sonnie murmured endearments. Pumpkin responded telepathically.

Turn a dog's paw upside down and rub the toes and palm pads. Dogs like foot massages, which promote good circulation in the paw and are relaxing. These thick pads are covered with sweat glands, so you are also rubbing the dog's scent onto yourself, which dogs want you to do. Dogs are remarkable animals that have, more than most, the gene for interspecies affection. It's something to think about as you rub their feet.

Trackers label toes as T1, T2, T3, T4, and T5, with T1 on the "inside" of the track like your thumb and big toe. Domestic dog tracks usually register four toes and heavy, blunt claws. A reduced T1 is higher up on the front legs only and can show when the animal is running. A heel or carpal pad is also higher up on the front legs. The length of a track—measured from claws to the bottom of the palm pad—ranges from less than one inch to more than four inches.

Pads are layers of tough skin over connective tissue that protect bones. In many tracking books, the pads on the toes are called digital pads. The palm pads on front feet are metacarpal pads and the palm pads on hind feet are metatarsal pads. Heel or carpal pads are sometimes referred to as proximal pads. The terms *metacarpal* and *metatarsal* also refer to the bones of the feet. Plantigrade animals like humans walk on all the bones of their feet. Digitigrade animals like dogs evolved from plantigrade animals in order to move faster. Both dogs and cats walk on their toes (or phalanges) and the end points of the metacarpal and metatarsal bones. The remaining bones of the feet moved upward, lengthening the leg and increasing speed.

When you know if a foot is front or hind, you might also use the terms *metacarpal pad* or *metatarsal pad*. Commonly, though, you will just say *palm pad*. This is what I have chosen to do, even when I am specifically describing a front or hind. I will also more often say *heel* rather than *carpal* for plantigrade animals whose tracks commonly show heels.

In all mammals, front feet are different from hind feet. Although this makes sense—the front of the body is different from the rear of the body—the result is a confusion of detail for the wildlife tracker. One easy difference to remember is size. A domestic dog's front feet are typically larger than the hind feet because most of the dog's weight is in the front. That principle holds true for any animal. Heavy head and shoulders (a coyote or cougar) mean larger front feet. Heavy thighs and rear (a bear or beaver) mean larger hind feet. Overweight dogs do not always show this difference between front and hind tracks.

Repetition is useful when learning something new, and I will deliberately repeat some information in these boxes. Here goes:

Domestic dog tracks are symmetrical. One half looks roughly like the other half. A rough X, or sometimes an H, can be drawn in the negative space without bisecting the palm pad. There is often a raised mound in the center of that space. The palm pads are small relative to the size of the track. Trackers like to measure this relative size by guessing how

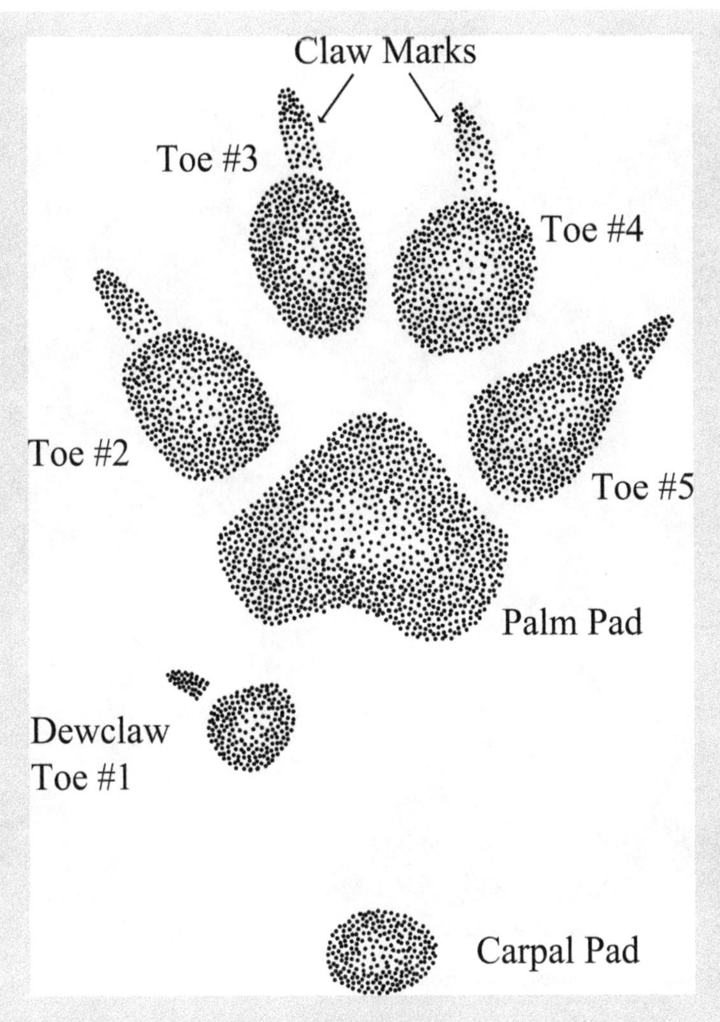

FIGURE 2.3 Domestic dog, right front

many toes would fit jigsaw-style into the area of a palm pad. The palm pads of wild canines hold between one and two and a half toes, depending on the species, but some domestic dogs like German shepherds can have much larger palm pads. The top of the palm pad has one lobe or a single point and the bottom of the palm pad has two or sometimes three.

Even in blurred and partial prints, the four toes, blunt claws, symmetry, relatively small palm pad, and X or H in the negative space can indicate that this is a domestic dog.

FIGURE 2.4 Domestic dog, front

The details of tracking involve geometry. In the front feet of a domestic dog, palm pads are triangular. In the hind feet, palm pads register as triangular to oval. The two middle toes are generally oval. Side toes can be oval to triangular. Serious trackers memorize this kind of information—and much more—for each species.

WILDLIFE TRACK AND SIGN CONVERSATION #3

"Domestic dog."

"Domestic dog."

"Domestic dog."

"Bird."

"Heron?"

"Right here by the river."

"And this? Oh, wait!'

"Tail drag! No tracks."

"Beaver."

"Slid right into the water."

"Brilliant."

"Domestic dog."

"Domestic dog."

"Domestic dog."

3

Coyotes Are the Original Aikido

Two other species of canines live in my neighborhood, just as present as domestic dogs but much less seen. They are waiting in the shrubs and grass, watching me walk or run or bicycle, flash, flash, flash, and lying low until that silliness is over. Then they trot across the road, sniff for mice, and consider a meal of mesquite or juniper berry. I will see their tracks, too, if I'm looking for them.

Adult coyotes weigh between fifteen and fifty pounds, the animals smaller in hot climates than in cold. Their muzzles are long and pointed, ears prominent and pointed, yellow eyes pointed slightly down. Despite the thick fur and bushy tail, the gestalt is angular, an animal slicing through the world. These omnivores mostly eat deer mice, woodrats, squirrels, rabbits, quail, scavenged deer or other animals, but also fruits, berries, and seeds. They have strong family ties, the male and female raising pups together. They are intelligent, curious, vocal, and versatile, moving easily in packs, pairs, or alone. They survived all those Paleolithic extinctions. They endure. There's something about them that reminds us of us.

For thousands of years, as soon as humans got to North America, we saw this laughing glimpse of a figure in our peripheral vision. We made up stories, hundreds of Coyote tales in dozens of Native American tribes. There are so many stories that author Dan Flores believes Coyotism was a kind of Paleolithic pan-religion. In *Coyote America,* he writes, "No other

FIGURE 3.1 Coyote
(photo by Elroy Limmer)

deity in America came anywhere close to inspiring such a vast body of oral literature."[1]

Coyote, the god, often got involved in the creation of the world, sometimes heroically, sometimes as a comic fool. He released Buffalo from underground, found Salmon a home, defeated Gila Monster. He brought fire to people. He also brought death. He juggled his eyes and lost them. He talked to his feces, and his feces talked back. He could behave badly, having sex with his daughter, having sex with almost anything—holes in the ground, vaginas in trees—his penis detachable as well as long. He murdered his friend Wolf and urinated on Woodrat's store of food. He is reviled for being ridiculous and extreme, and he is cherished for being ridiculous and extreme. As a god, he died many terrible deaths. He was continually reborn.

Prophetically, this last part came true. Anglo-American ranchers and farmers have long waged war against animals who threaten their livestock, and the coyote's response to newborn calves and lambs seems to have been both astonishing and outrageous. Thirty years ago, I wrote a book about grazing on the public lands of the American West. The cruelty of that war startled me then, when I was younger: coyotes torn apart in medieval torture scenes, pups burned alive in their dens. A common form of death was the M-44, an ejector device loaded with sodium cyanide and driven into the earth. When a coyote (or domestic dog or other animal) pulled on the bait positioned aboveground, a spring-activated plunger propelled the poison into its mouth.

Most North Americans kill coyotes not for meat or fur but as a pest, something noxious and offensive they want to get rid of. Killing coyotes remains a sport in parts of America, with contests and cash prizes for the highest number of bodies brought in. At the 2020 Eastern U.S. Predator Calling Championship in Virginia, officials awarded a rapid-fire, magazine-fed assault rifle for the largest coyote (45 pounds) and another for the smallest (17 pounds). In two days, 213 teams killed 850 coyotes and foxes. Children are encouraged to participate in these events, with the last traditional Santa Slay Coyote Tournament held in Arizona in 2019.

At the same time, the larger public reaction is shifting rapidly. Certain behaviors cannot survive YouTube. Arizona, along with a handful of other states, has since outlawed wildlife killing contests. The National Coalition to End Killing Contests includes dozens of groups, from the Humane Society Veterinary Medical Association to Great Old Broads for Wilderness.

The service of killing "pests" is something the government has provided since the Animal Damage Control Act of 1931, which authorized "the destruction of mountain lions, wolves, coyotes, bobcats, prairie dogs, gophers, ground squirrels, jackrabbits, and other animals injurious to agriculture, horticulture, husbandry, game, or domestic animals, or that carried disease." A government agency called Wildlife Services remains officially in charge of killing coyotes—an average of more than 60,000 each year for the past five years. Although that number seems high, the number of coyotes killed unofficially is much higher, perhaps another half million a year.

Almost everywhere in North America, you can kill a coyote on private land at any time. On public land, though, you need to follow certain rules. Wildlife management varies from state to state in America and province to province in Canada, and the differences can be extreme: an animal hunted or trapped in one state might be on the threatened or endangered list of another. This makes sense when a species is at the edge of its habitat or in decline in one area but not in another. Sometimes, however, the difference is purely political, or rather emotional—what people perceive or feel about an animal regardless of its status and biology.

In most American states and in much of Canada, the rules for coyotes on public land are simple and mimic those for private land: coyotes can be hunted or trapped year-round without any restriction as to number, sex, or age.

As a god, Coyote is almost always male. As a survivor, it's all about the female. A typical pack consists of a mated or alpha pair, their offspring from the current year, and some juveniles from the previous year. Females are receptive for only a few days sometime between January and March, with a gestation of about two months. In a litter of four to eight pups, perhaps half will die. Both the alpha female and male raise the pups, with nonbreeding juveniles helping. The younger females may never breed. Daughters may stay with their parents their entire lives. In this and other ways, coyotes stabilize their populations to match surrounding resources of food.

But coyotes have a strong, a very strong, "responsive reproduction." When their populations are threatened and in decline, more females breed, younger females breed, and more pups are born in a litter. With fewer coyotes in an area, more food might be available, so more pups in the litter survive. When packs break up or are destroyed, coyotes have the ability to live in smaller groups or as individuals. Coyotes are the original aikido. Persecuted and despised, they have used our energy against us, expanding and flourishing in areas once inhabited by wolves, which survived less well the twentieth century's assault of poison, traps, and federally funded hunting. We don't really know how many coyotes are in the United States. As with many wildlife populations, we assume that mortality is proportional to abundance. If people are finding a lot of coyotes to kill, then probably there are a lot of coyotes available to kill. We kill a half million coyotes a year and seem to have more than ever.

Coyotes travel easily. Once in a new territory, they can breed with other canines. In North America, eastern coyotes have slightly larger tracks than western coyotes, with the same compact shape and show of sharp claws. An analysis of twenty-five studies estimates that northeastern coyotes have a genetic makeup of roughly 60 percent coyote, 30 percent wolf, and

10 percent domestic dog.[2] The mixing with wolves probably began more than a hundred years ago when western coyotes entered the Great Lakes area, where a decimated wolf population was desperate for mates. Feral dogs were introduced into the mix about fifty years ago.

Recently, there has been little or no mixing of coyotes, wolves, and dogs. Some scientists believe the eastern coyote should be considered a new species since that genetic mix now seems stable. Eastern coyotes in southern states, however, show a slightly different genetic mix, and the entire eastern subspecies is sometimes called "coyote soup." Eastern coyotes may be larger, with small changes in teeth, but they do not behave much differently from western coyotes.

For decades now, coyotes have been moving into cities and suburbs, where life is easier and safer than in the country. They continue to feed on rats, mice, squirrels, rabbits, birds, and deer, as well as roadkill, fresh fruits, nuts, and seeds. They shelter under buildings and in bushes and the odd landscaping feature. They den in parks, railroad yards, gravel pits, and dirt lots. They crowd together more, with smaller territories. They live longer, if they can avoid traffic. They are working on that. Researchers in Chicago have watched the movements of radio-collared coyotes who wait at stoplights and cross when the intersection is free. On a divided highway, these coyotes look in the direction of oncoming cars before crossing to the median. Then they look the other way.[3]

City coyotes are a reminder that cities are part of the natural world and that we are part of the natural world. The natural world is messy—astonishing and outrageous from the human point of view. Pets are estimated to be a very small percent of a city coyote's diet, although that doesn't mean much when it's your pet. Coyotes will also kill unprotected pets for territorial reasons. Aggressive behavior toward humans is uncommon and almost always related to a den close by. But coyotes in neighborhoods and suburbia can become less frightened of humans. Very rarely, coyotes do attack people, with two recorded deaths in the past fifty years. Both were tragic, a three-year-old girl in California and a nineteen-year-old woman hiking in Nova Scotia.[4]

Most problems with coyotes result from people feeding them. In 2009, in a large park in Los Angeles, a man lying on the grass was unexpectedly nipped on the foot by a coyote. The coyote then trotted to a nearby spot, sat, and waited. This was the animal's notification behavior. Other people in the park had been giving out food, and the bite was a reminder that dinner was late. A few days later, a woman had a similar experience. As a consequence, the Department of Agriculture sent a team into the park that killed all the coyotes—seven—they could find.

Wildlife managers say the best way to treat urban coyotes is to harass them. Yell, bang a pot, make a shooing motion. You don't have to mean it. Your heart might lift at the sight of that yellow-furred Coyote god, but pretending otherwise creates a safe distance for you and for him.

What we especially love about coyotes, what we see reflected—something about us—is their ability and need to communicate. Coyotes growl, huff, woof, bark, and bark-howl as a warning or sound of alarm. They whine, yelp, woo-oo-woo in greeting and submission, saying "Hello," saying "Be nice to me." Famously, coyotes bark-yip-yip-yip-howl alone or as a group, with individuals easily recognized by their habits of timbre, cadence, and warbling. In a pack, alpha pairs usually start the howl, other members join in, and nearby packs respond. This kind of choral song is happening all over North America, although you have to be outside and alert to hear it.

Barks and howls can also convey distance. These performances help coyotes know where another group is and who they are. Sometimes sound is deceptive, and deliberately so. Coyotes can seem like a larger pack with wavering howls and rapid changes in pitch, amplified by echoes bouncing off rocks and trees. Mostly, though, questions are raised and answered. "How did you get so lost?" "Come home quickly." "Wait, stop. Don't come over here at all." "Whoa, listen to how many of us are competing for resources! Let's just breed once this year." And within members of the same group: "We're amazing! How good we sound, how strong, how united."

In the small town of Silver City, New Mexico, I am in a community choir, a beta female, not very good at singing, just happy to be part of the pack. We choose sweet songs, "The Extraordinary Light of Your Being," "White Coral Bells," "We Hold You in Our Circle," for their lyrics about solidarity and optimism. Each time, once a week, I feel the joy of raising my voice with others, the satisfaction of harmony. Music carries me out of my pain or loneliness and creates in that moment a purpose and meaning. It seems odd to think coyotes don't feel this, too.

I am always pleased to see a coyote track. The shape is oval. The X is clear. Typical for a wild canine as opposed to a domestic dog, the four toes are relatively close together and pointing forward. Trackers would say the print

FIGURE 3.2 Coyote, front and hind

is compact, with little splay. These trackers carry rulers, but I often forget mine. Finding some part of your body to measure with is handy. (A Coyote story would make a joke about that.) I use my first or index finger. From the tip of that finger to the crease of my first joint is about one inch. From the tip of that finger to the crease of my second joint, about two inches. From the tip of that finger to the base of my finger, about three inches. A coyote track here in the American Southwest would measure roughly between the crease of my second finger joint and the base of my finger.

Along my dirt road in southwestern New Mexico, I am not going to confuse this print with those of Michelle's German shepherds, which are so much larger, with splayed toes and blunt heavy claw marks. But some dog tracks are similar in size and shape, and you can't always tell the difference. One clue is that domestic dogs tend to leave a heavier imprint than coyotes, and their tracks are relatively flat. A coyote walks first on its toes so that the track is angled, palm pads registering higher than toes.

Another difference might be how the animal is traveling. Domestic dogs are usually well fed, with energy to spare, so they meander. They lark. They goof. Coyotes are more cost-effective, getting straight to where they want to go. Often they place a hind foot over the front foot in what is called a double or direct register. Coyotes typically leave a straight line of double register tracks when they are trotting, their favorite way of moving through the world. Coyotes also walk and trot in each other's footsteps, so that what seems like one coyote may actually be two or more.[5] Purpose, efficiency, caution—a single track can emanate these qualities.

I've sensed that other presence, too. I've squatted before the print of a coyote and felt the brush of something against my arm. Laughing Coyote. Old Man Coyote. How can you not believe in gods this particular afternoon? The wolfberry glows. Clouds race across the sky. There's the chalky smell of dust. Delightful existence. What could be taboo? Certainly not the stuff of matter, gristle and rock, mucous and feces, penis and vagina. It's all good. It's all Creation.

Coyote tracks usually register four toes. A reduced T1 and heel pad are higher up on the front legs. Small claw marks can look like triangles or circles. You might see only the two claw marks of the middle toes. The front track of an adult coyote in the West—measured from claws to the

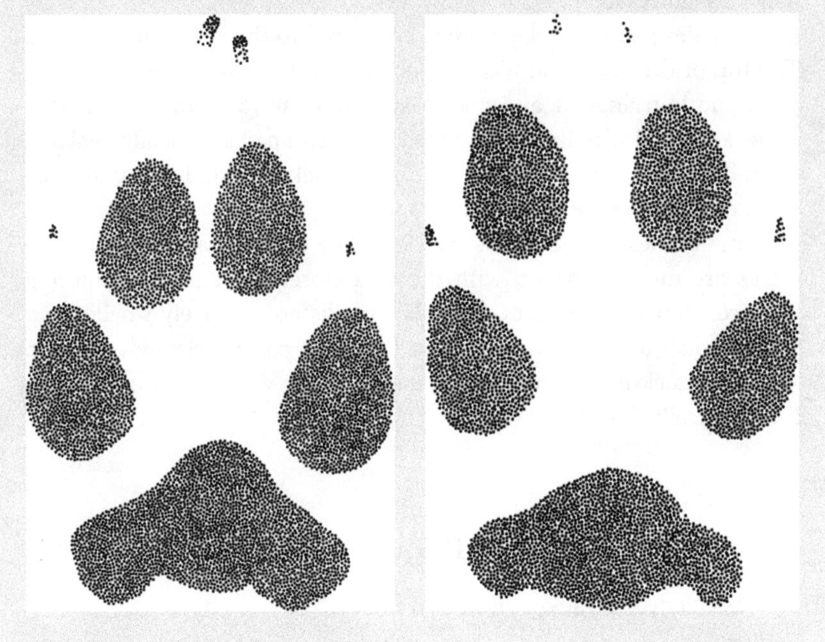

FIGURE 3.3 Coyote, front and hind

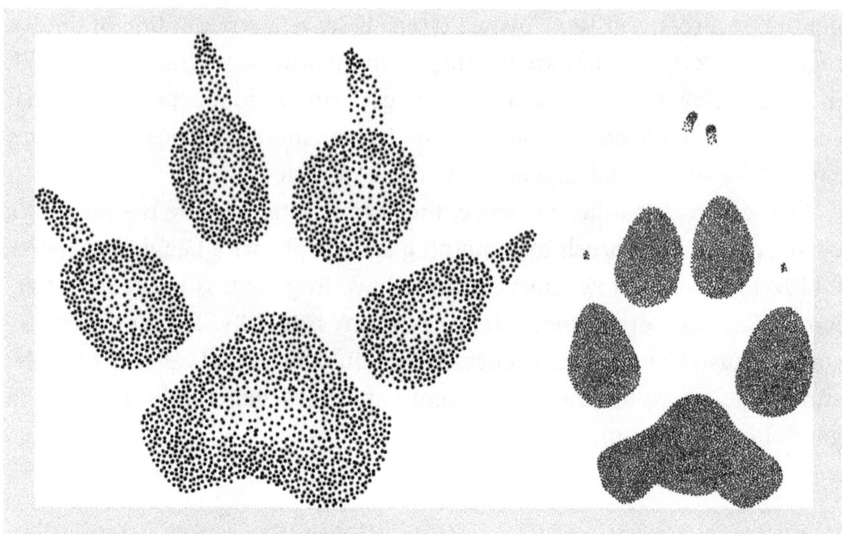

FIGURE 3.4 Domestic dog and coyote

bottom of the palm pad—is slightly over two inches to slightly over three inches, with the hind tracks somewhat smaller. Eastern coyote tracks may be slightly larger.

The palm pad is relatively small compared to the shape of the track. The top of the palm pad has one lobe or point. In the front track, the palm pad is triangular and may have two small lobes or "wings" that project out or down. In the hind track, the palm pad is typically oval and often has small lobes on either side. The track is roughly symmetrical. An X can be drawn in the negative space.

Some domestic dog tracks can look like coyote tracks. But coyote tracks are more compact, with the toes closer together and pointing forward, showing less splay. The claw marks are relatively smaller and sharper than a domestic dog's, and the palm pad is relatively smaller. A coyote track might show the presence of fur while a domestic dog's track typically does not.

WILDLIFE TRACK AND SIGN CONVERSATION #45

"Hey, classic coyote track. About two inches."

"Mmm."

"There's the X in the negative space formed by toe pads and palm pad."

"Mmm."

"And I can see claw marks."

"Mmm."

"The toes are compact, close together, so I don't think this is a domestic dog. Why are you saying *Mmm*?"

"Hmm."

"Hmm what?"

"Sometimes a lagomorph leaves an impression in a track which looks like an X. Of course, lagomorphs have claws, too."

"Lagomorph, as in . . ."

"Black-tailed jackrabbit."

Silence.

"I love black-tailed jackrabbits."

"So big, so virile. Such zigzags. Such leaps."

"They make me think of *Watership Down*."

"I loved that book."

4

Let's Just Enjoy Foxes

The artificial ranking of taxonomy is kingdom, phylum, class, order, family, genus, species. The mnemonic I learned in high school was King Phillip Came Over for Good Soup. Eventually, in college, King Phillip Came Over for Great Sex. Sometimes he came over for spaghetti. The kingdom Animalia includes the phylum Chordata, animals with backbones. The phylum Chordata includes the class Mammalia, hairy, warm-blooded animals that nurse their young. The class Mammalia includes the order Carnivora, typically animals with paired upper and lower teeth called carnassials that allow for the shearing of flesh. The order Carnivora includes the family Canidae. The family Canidae includes the subfamily of our modern Caninae, divided into genus and species.

Gray foxes, among the first caniforms to diverge from the order of carnivores, kept some of their feliform abilities, notably semi-retractable claws. Claws that are not in constant use are sharper and good for climbing trees. Gray foxes also retained the catlike ability to rotate their forearms, which allows them to hug a trunk as they push upward with their hind legs. With relatively short legs and a low center of gravity, they can scramble sixty feet up a cottonwood and then jump from branch to branch. I've waited all my life to see this.

Most often active from dusk to dawn, gray foxes have black-tipped tails, black markings under their eyes, and coarse salt-and-pepper coats.

FIGURE 4.1 Gray fox
(photo by Elroy Limmer)

Adults weigh seven to sixteen pounds and blend easily into dense cover. More than other canines, they can digest grains, nuts, and fruit, which is why the first European settlers shot them from the trees for eating their crops. As more forests were cut down in the East, the gray fox may have been outcompeted by the North American red fox, which is in a different genus and does not climb trees. Today, both species range through much of the United States, with the gray fox extending only barely into Canada (they are not great in snow) and south into Mexico and parts of South America.

As with coyotes, we don't really know how many gray foxes are in the world. The International Union for Conservation of Nature (IUCN) is an international group that monitors wildlife and evaluates the risk of extinction. In 1964, the IUCN established its Red List of Threatened Species, drawing from the most recent scientific research. The nine categories of the Red List are the following: not evaluated, data deficient, least concern, near threatened, vulnerable, endangered, critically endangered, extinct in the wild, and extinct. When they can, the IUCN makes an educated guess at the worldwide population of each species and the population trend—increasing, stable, or decreasing.

The IUCN is not perfect. Their reports depend on what scientists are researching what animals and what governments are reporting what kinds of information. The data will not always be updated, accurate, or free of politics. But the IUCN is what we have—a network of some 15,000 experts doing their best with what they have.

The IUCN lists the coyote as a species of least concern, with a population trend of increasing. For the gray fox, the IUCN also says least concern, with a population trend of stable.

In the United States, as with coyotes, we guess that gray foxes are doing well based on how many were reported killed the previous year. One estimate from the Arizona Sonoran Desert Museum puts that at half a million. Gray foxes are mostly trapped and hunted for their fur, not as pests. We might also factor in how often these animals breed and how much habitat they have. We include any information we have about disease or deaths by car. We may use reports from people who simply see gray foxes. Then we kind of squint and say, "Lookin' good."

I have been using the word *we* to represent those of us who live with gray foxes. But the actual decisions about how to manage gray foxes are made by a select handful of people.

Every state in the United States has a wildlife agency that stands alone or is part of a natural resources or environment department. In most states, the governor appoints a commission or board that oversees the work of this agency. Typically, this oversight includes setting policy for budget, for wildlife conservation, for education, and for hunting, fishing, and trapping. Many states require that a percentage of their commissioners represent the interests of hunters and fishers, as well as farmers and ranchers. Even in states without that requirement, most wildlife commissioners traditionally self-identify as a "consumptive user." In 2023, for example, the twenty-one commissioners of Wyoming, Idaho, and Montana were 100 percent non-Hispanic white, 81 percent male, and 91 percent self-designated hunters or fishermen. This might be compared to the 4.6 percent of people in America who own a hunting license and the 17 percent who fish.

If the worldview of hunting and fishing dominates wildlife management, that's partly because our wildlife agencies are funded with the fees and licenses paid by hunters, fishers, and trappers, as well as federal taxes paid on all guns, bullets, and archery equipment. For many state agencies, these sources are more than two-thirds of their revenue. Bullets actually bring in the most money—although notably most guns and bullets are not bought by hunters. Federal grants based on the sales tax on weapons and ammunition must have a certain amount matched by the state. Wildlife agencies not only scramble for money to match federal aid but also have an incentive to increase hunting and gun sales.

Superseding state policy are federal laws such as the 1918 Migratory Bird Treaty Act, which protects more than a thousand bird species, and the 1973 Endangered Species Act, which identifies and protects animal and plant species threatened with extinction.

In Canada, the situation is similar. Most of the authority for the management of wildlife lies with the thirteen provinces and territories, with the federal government having more oversight over endangered or threatened species. Canadian wildlife agencies are also strongly tied to consumptive use, even though a relatively small percentage of Canadians hunt. The funding for wildlife in Canada, however, comes mainly from general tax revenues, not any special interest group. This could be the future of America, too, with the passage of new bills and acts that provide federal money to states for habitat restoration and the recovery of nongame species.

You have guessed by now that I am not a hunter or trapper. This is by choice rather than missed opportunity. However, I do eat meat, both hunted animals and factory-farmed. My husband and friends hunt deer and elk for food, and I respect hunting as a lifestyle, a meaningful recreation, or a spiritual practice. Killing prey animals like deer can be necessary to reduce populations that are damaging ecosystems. Killing predators that threaten human life is also necessary. In short, I accept death as natural and essential. Animals kill each other. Animals die. Even you and I. Obviously, if I am going to write about wildlife in North America, I am going to write about hunting and trapping. I am going to treat hunters and trappers with the empathy and intelligence I want them to treat me. As best I can, I am going to work through the nuances and contradictions of my own belief system as well as some of the beliefs that shape wildlife management in North America.

Even so, I'll save the discussion about fur trapping for the chapter on bobcats. Let's just enjoy foxes.

Gray foxes raise their pups together and probably mate with the same partner every spring. They can live six years in the wild but often die before that, being susceptible to outbreaks of diseases such as distemper and mange. In southwestern New Mexico, we have seen their populations crash because of rabies. We mourn—where are all the foxes?—even though we really don't want to see a rabid fox. As well as being trapped for their fur by humans, gray foxes are also hunted and killed by other animals such as coyotes and bobcats, which dislike them as competitors and threats to their young. In one study in California, eight of twenty-four radio-collared gray foxes were killed by coyotes and two by bobcats. In the Sonoran Desert, one biologist estimated that coyotes accounted for 90 percent of gray foxes killed, with a coyote pack typically ambushing the fox, pinning it to the ground, biting its neck, and then not eating the body.[1] Owls, eagles, and mountain lions also kill gray foxes, which in turn are incessantly hunting and killing squirrels, rats, mice, birds, and large insects.

Partly as a way to avoid coyotes and bobcats, gray foxes sometimes rear their pups close to human houses. Over the years, I have watched these families from the window of the room where I write. These are glimpses only. Two foxes sunning on the slope that leads to the irrigation ditch, one lifting its leg to lick its belly. A fox racing by the bird feeder, rat tail dangling from its mouth like a film noir cigarette. A fox being schooled by a skunk. This fox, I assumed, was an adolescent half-interested in the skunk as prey or playmate. The young fox came closer and closer until the skunk, which I imagined to be an older female, stomped once with her front feet. Up on hind legs, then down on the ground. She also emitted a slight odor. The fox catapulted. Just the right verb. Now the fox was a dozen feet away, looking surprised.

The gray foxes I have glimpsed over the years come down a tree-filled, leaf-littered, branch-strewn incline from the irrigation ditch that runs through our small acreage. Between this dirt ditch and the paved country road is another acre, a thicket of four-winged saltbush and honey mesquite. All this is perfect for mice and rats and gray foxes. Perhaps these denning foxes have dug a hole in the ground or expanded the cavity of a fallen cottonwood. I have never been tempted to find these dens, not even when I suspect there might be pups inside, napping in an adorable tangle. If I am lucky during the months after their birth in April or May, I'll see the maturing pups briefly, until by seven months they are fully grown and a few months later ready to leave. I never get any closer to these animals. I've never been inside their homes. I live in the rural West and have plenty of neighbors like this. Our politics are different, our beliefs and worldview. Sometimes that's hard to bear, but we each make the effort to be courteous. In the case of foxes, if I ventured too close, they would move their pups.

These vignettes are from years of reaching for my binoculars. Some years, I never see gray foxes at all. Naturally, I have many memories from these last decades. Presidential elections, holidays with adult children, walks with friends, students, museums, movies, dancing parties. The gray foxes rank similarly in importance, helping create the story I tell myself about myself. I am a person who gets to see gray foxes. I live in a world of foxes unaware of my interest in them, as are the skunks. Our relationship is not personal or transactional, not about me at all.

In a typical gray fox track, about one and a half to less than two inches, four toes point up with the compact quality of a wild canine. Often you see the impression of fur in the track. Fur accounts for the relatively large negative space in the center. You can draw an X in the negative space between toes and palm pad—except that in the front track of a gray fox, this negative space might look more like an H. Also, because the sharp curved nails of a gray fox are semi-retractable, sometimes the claw marks do not show. Sometimes, especially for incomplete tracks, a gray fox track is mistaken for the round clawless track of a large domestic cat or a small bobcat, with the feline's distinctive three posterior lobes in the palm pad. In mud, too, all tracks tend to look splayed and not compact as the animal spreads its paws for a better grip.

I'm sorry about this. I'm especially sorry about incomplete tracks because there are so many of them. I'm sorry to say that coyotes, obviously, like to play and lark as well as be purposeful and efficient, so sometimes their tracks meander like a domestic dog's. The general rules about tracking are beloved and treasured, like old family stories. But they are not really to be trusted. X for canines but sometimes an H. Claws for canines but sometimes not for foxes.

The most abundant and best known of North American foxes is the American red fox, which ranges throughout Canada and much of the United States. Red foxes are good swimmers and agile hunters, with a distinctive pouncing leap like a diver into a pool—only into grass. Some people criticize these animals for their surplus killing when, excited by too much prey, they ravage a chicken coop or a nesting colony of seabirds. To be fair, they will cache at least some of that meat for later. Red foxes also

FIGURE 4.2 Gray fox, left hind and right front

impress opponents by turning sideways, arching their back, and raising the hair along their spine. The tunnels in their dens can be seventy-two feet long. They snack on grasshoppers. They have all kinds of interesting natural behaviors. For this, I highly recommend Mark Elbroch and Kurt Rinehart's *Behavior of North American Mammals*.

Mostly, though, we know red foxes for their fur, gleaming white at the throat and breast, contrasting with the rest of the coat, which is orange-red, hot-pepper-red, ginger, lava. Red foxes can also have mottled coats of brown and gray or patches of brown or gray. But those aren't the ones you see photographed.

Like the coyote, like us—generalization, generalization, generalization—red foxes are both a synanthropic species, benefiting from human-altered environments, and an edge species, occurring in the borders between habitats.

FIGURE 4.3 American red fox
(photo by Elroy Limmer)

Naturally, they have adapted to our cities and suburbs. Since cities provide more food, one behavior change is that animals who usually compete for resources are now more relaxed. Researchers in Madison, Wisconsin, report that coyotes and red foxes have been seen hunting close to each other without aggression or fear. In one case, a male red fox and male coyote nosed about a park, separated by only twenty yards, without the coyote attacking or the red fox fleeing. City foxes also seem less concerned about moving their pups from dens when city coyotes are nearby.[2]

The IUCN does not guess at the population of American red foxes, and we all assume they are doing well. This is not true, however, of red fox subspecies struggling with global warming in the forests of the Cascade Range, the Sierra Nevada, and the Rocky Mountains. Importantly, subspecies are at a granular level the IUCN does not address. Monitoring the health of these subspecies is the job of government agencies, conservation groups, scientists, and everyone else who lives in community with red foxes.

Four other species of fox are in North America, and they, too, are not thriving, despite an IUCN listing of least concern. The arctic fox is similarly challenged by global warming, with new diseases, a decline in its prey of lemmings, and competition with the encroaching American red fox. The swift fox, habituated to grasslands, has been listed as federally endangered in the United

States and Canada but is recovering thanks to heroic efforts. Probably fewer than a thousand swift foxes now live on 40 percent of their historic range. The island fox of the Channel Islands off California suffered a catastrophic decline in the 1990s and was listed as federally endangered but has since recovered to more than two thousand animals. The kit fox lives primarily in the American Southwest and Mexico. Its populations are low or unknown.

Like the gray fox, the kit fox may be somewhere around my house, although I've never seen one. This small fox, three to six pounds, has a light coloring that blends easily into the desert landscape. The kit fox also hunts at night and retreats during the day into an elaborate den with many entrances. The animal is considered locally endangered in parts of the United States.

IUCN FOX ENDANGERED STATUSES

Gray fox, least concern, population trend stable
Red fox, least concern, population trend stable
Arctic fox, least concern, population trend stable
Swift fox, least concern, population trend stable
Kit fox, least concern, population trend decreasing
Island fox, near threatened, population trend increasing

All foxes have a special prideful relationship to their urine and feces. They urinate often in dribbling amounts, leaving their scent to mark territory, as memos to self about food caches, and as sexual advertisement. Red foxes have been known to drip and drop as much as seventy times an hour while foraging, a constant writing and reading in scent. Red fox urine is potent, and trackers have reported scent "clouds" wafting through the forest.

Urine is basic communication. Feces makes a more powerful statement about your relationship to the world. Like many predators, foxes defecate on top of the feces left by other predators—maybe, especially, those coyotes and bobcats. Foxes like to defecate on top of anything elevated—rocks, stumps, walls—so that the smell will better waft through the air and their statement be clearer and more impactful. Foxes like to leave their feces in the middle of a trail or road. The gray foxes who live near my house like to leave their feces on my porch. In all my years living here, I've glimpsed gray foxes a few dozen times. But I've seen hundreds and hundreds of examples of their feces, sometimes orange and red, mushy and speckled with digested fruit, sometimes black from a purely carnivorous meal, a twisted tube folded onto itself.

In the tracking world, feces are called scat. The details of scat are almost as complex as those of tracks, with more numbers to remember and more exceptions to the rules. I don't know most of these details, and I am not promising to learn them. What I do experience as I get on my knees, find a twig, and pull apart scat is the visceral recognition of bone—this omnivore ate a mouse. I recognize fur. Once a rabbit. The absence or presence of seeds. The moist dark of organs. The gestalt of a predator. We digest each other. It hits you in the stomach. The violence and beauty of the food chain.

It's also nice (and sometimes presumptuous) to tell the difference, say, between deer droppings and elk droppings. Just to comment, "Too large for deer" or "Look at that oval shape." I am especially fond of lizard scat, which is so distinctive, a dark pellet with a cap of white uric acid. I impress myself when I note that a J-shaped turkey scat is from a male, and a coiled globby turkey scat is from a female. Like many birds, turkeys expel from the cloaca, which is the site of their reproductive organs. Their scat reflects the different shape of those organs in males and females. You either get a thrill out of knowing this, or you don't.

Gray fox tracks usually register four toes. A reduced T1 and heel pad are higher up on the front legs. Claw marks are very fine and often do not show. The front track of an adult, measured from claw marks to the bottom of the palm pad, is roughly one and a half to slightly under two inches. The top of the palm pad on the front foot has one lobe or point while the bottom of the palm pad may have three lobes or a distinct wavy pattern. The negative space may have the shape of an H. Front tracks are larger and rounder than hinds. In the hind track, the side toes are often tucked under and behind the middle toes. The negative space forms an X. The palm pad may register as a circle or a circle with small lobes or wings on either side. Both front and hind tracks may show signs of fur.

Red foxes are slightly larger than gray foxes, as are their feet and tracks, reaching almost three inches in the front and two and a half inches in the hind. These feet are also furrier and better in snow, blurring the print of toes and palm pad, with more negative space in the track. The front palm pad can appear in the shape of a chevron or flat line. The sharp claws are semi-retractable but often present.

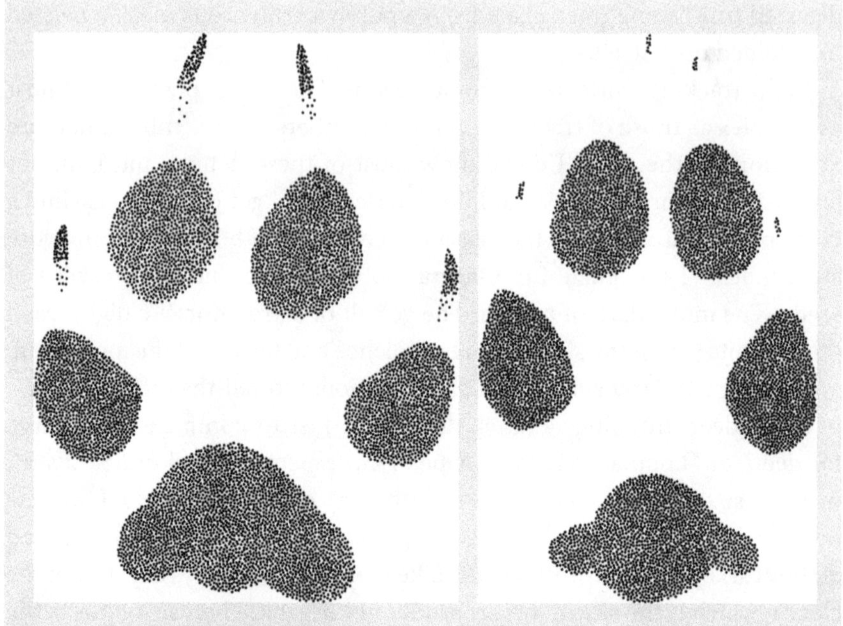

FIGURE 4.4 Gray fox, front and hind

FIGURE 4.5 Gray fox, front and hind

The tracks of a red fox are larger than those of a gray fox but smaller than those of a coyote. One of the best books for these details, Mark Elbroch's *Mammal Tracks and Sign: A Guide to North American Species*, second edition, with contributions by Casey McFarland, gives very exact

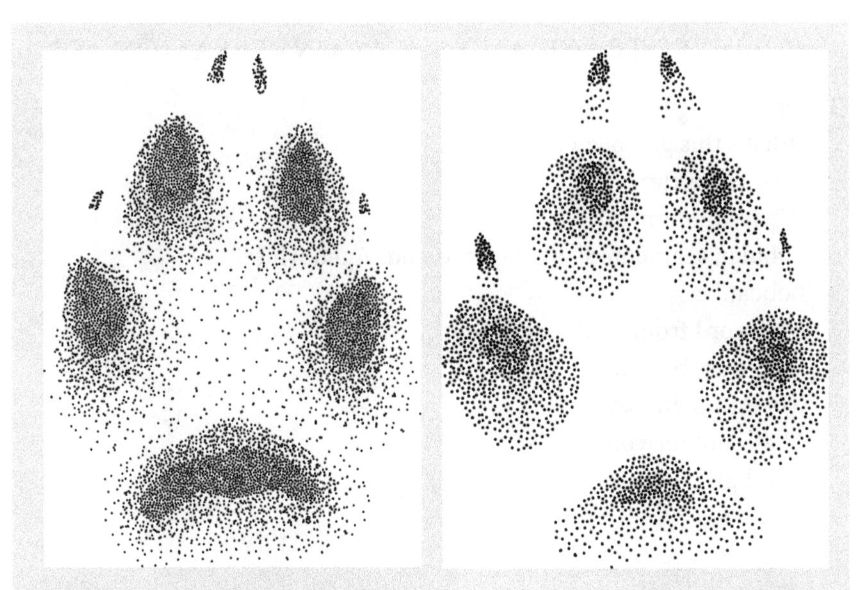

FIGURE 4.6 Red fox, front and hind

dimensions. Measuring from claw marks to the bottom of the palm pad, front tracks in a red fox are 1⅞ to 2⅞ inches with a width of 1⅜ to 2⅛ inches. Hind tracks are 1⅝ to 2½ inches long with a width of 1¼ to 1⅞ inches. Measurements are also given in centimeters in the hope that someday Americans will be as smart as the rest of the world.

Determining what animal made that print in front of you can revolve around such measurements. You'll decide coyote, not dog, or bobcat, not mountain lion, based on size. For my purposes, however, I use terms like *roughly* or *about an inch* and round up to the nearest half inch any measurements taken from various tracking sources. Measuring precisely in the field is problematic for beginning trackers, who are still trying to find that bottom of the palm pad, especially in partial or blurred tracks. Measuring precisely is complicated by the substrate of the track, whether it's mud or soft dirt. Measurement is only one factor, too, in what one of my mentors, Bob Ollerton, calls "multifactor analysis." Don't depend on any one thing. Look for the gestalt. The many things.

WILDLIFE TRACK AND SIGN CONVERSATION #67

"Fox scat."

"What's this guy been eating?"

"Juniper berries."

"What's underneath?"

"Tubular, segmented, one tapered end. Aged white."

"Bobcat."

"Last word from fox."

"Take that, bobcat."

"No, Pumpkin. No!"

"That's not for you!"

"Good girl."

"Good dog."

5

It Almost Seems Wolves Should Be Allowed to Vote

Nothing says "I'm sorry" like reintroduction. We extirpate a species from its home, intentionally or unintentionally, and then spend a lot of money and time returning that species to its home. In North America, we have done this for black-footed ferrets, California condors, Pacific fishers, Texas horned lizards, Arctic graylings, California bighorn sheep, American flamingos, American bison, Hawaiian geese, musk ox, Blanding's turtles, Missouri river otters, golden eagles, bald eagles, and more.

Reluctantly, we have done this for wolves.

The Europeans who colonized North America had a special enmity for wolves. The country seemed full of them—one estimate is two million—just as the landscape seemed full of the food that wolves eat. This abundance represented an expansion of wildlife that had been building for years, ever since the earliest European explorers in the fifteenth and sixteenth centuries brought diseases like smallpox and measles that raged through the cities, towns, and villages of Native Americans. As many as 90 percent of indigenous people in the "New World" died, with current research suggesting that number to be at least fifty-six million in both North and South America.[1] In North America, with a population of some three to six million, people had long used fire to maintain grassland and sustain herds of grazing animals. In the relative absence of people, those herds swelled. Bison dominated. Elk, deer, and pronghorn flourished. So did wolves, until they were shot, trapped, and poisoned as competing predators, killed fervently at first, and then systematically.

Today we believe some sixty thousand gray wolves remain in Canada, where they are legally hunted, perhaps another ten thousand in Alaska, where they are also killed for fur and as predator control, and about six thousand in the contiguous United States. The IUCN considers them of least concern, with a stable population trend. Maybe two dozen critically endangered red wolves live in North Carolina. Fewer than five hundred eastern wolves—which the IUCN consider a subspecies of gray wolf—live mostly in Canada, where they are protected from hunting. We have such modestly specific numbers because of the wolf's endangered and threatened status in some parts of the United States. Also, wolves really interest us.

In 1995, gray wolves were reintroduced into Yellowstone National Park. From an ecological perspective, this has been a complete success. Wolves started a cascade. They ate the elk that had been overpopulating and degrading stream areas, which allowed willow and other vegetation to

FIGURE 5.1 Gray wolf
(photo by Elroy Limmer)

recover, which allowed beaver to spread and build dams, which stabilized streams by storing water for recharging water tables. The trees provided habitat for songbirds. The dams meant cold, shaded water for fish. Wolf kills became regular meals for scavengers that once relied on the seasonal deaths of elk in winter. Today, throughout the year, black bears, grizzly bears, coyotes, ravens, eagles, and magpies feed on wolf-killed carrion. One biologist added to that list of animals, "Beetles, wolverine, lynx, and more. I call it food for the masses."[2]

Wolves caused a significant decline in their natural prey of elk, although not mule deer or bison. They also killed coyotes and their pups, reducing by half this unusually dense population. Restoring the balance between wolves and coyotes had a surprising effect. Too many coyotes had meant fewer rodents, and when those voles and mice rebounded, so did owls and hawks. So did weasels and foxes. Coyotes feed on newborn pronghorn, but wolves seldom do, and suddenly there were more pronghorn.

Something else happened culturally in America. A few of the wolf packs in Yellowstone could be observed easily with spotting scopes, a technology that wildlife managers, scientists, and tourists began to enjoy. People watched wolves for hours a day, every day of the year, for years at a time. Some of the wolves were caught and radio-collared and their movements tracked. People took photographs and videos and wrote about their experiences. We were seeing into the lives of wolves, and those lives were so much more varied and emotional than we had thought. Wolves had their own stories and histories. This is something indigenous people already understood. It was new to the rest of us.

Wolf packs are family groups, and wolves don't like to mate with relatives. When an alpha male or female dies, the pack may need a new leader from outside the pack. In the late 1990s, a two-year-old black male known by researchers as 21 crept into Yellowstone's Lamar Valley hoping to woo an alpha female, known as 40, that had just lost her alpha and beta males. The alpha female 40 had assumed leadership by driving out her mother, and she often harassed her two sisters, 41 and 42, pinning them and forcing their submission. Alpha wolves do this on occasion, but not so often as 40 did. The aggressive 40 accepted 21 as her mate and for three years either kept her sisters from breeding with him or killed their pups.

In the spring of 2000, however, researchers could see that 40, 42, and one of 42's adult daughters had all mated with 21 and become pregnant. Three dens were now spaced across the pack's territory. Watching one evening at his scope, National Park Service employee and longtime wolf watcher Rick McIntyre saw the alpha female 40 come upon her sister 42 and another female and force them to submit. Then 40 headed over to 42's den of month-old pups, being babysat by a third female. By now, both the alpha male 21 and other male and female yearlings in the group were trying to feed all three litters, which meant less food for 40 and her young. As he writes in his book *The Reign of Wolf 21*, Rick "had a bad feeling" and "the theme song to the movie *Jaws* came to mind. . . . If she had killed her sister's pups once or twice before, she could do it again."[3]

The next morning, Rick and other researchers found a bleeding, frightened wolf near a Lamar Valley road. They expected to recognize the gray form of 42. But this was the tyrannical 40, bitten and mauled by more than one attacker. The humans tried to help, but the wolf soon died of her wounds. The subordinate females had rebelled and killed their alpha. After that, the researchers thought, 42 or her daughter would likely kill the alpha's pups or let them starve.

That didn't happen. Instead, 42 went to her sister's den and nursed those pups. Then she moved all her pups, one by one, to the main den, swimming the Lamar River and crossing a major highway. Her daughter moved her litter as well, creating a giant nursery of twenty-one pups. Twenty-one! Five-year-old alpha male 21 clearly worked the hardest now feeding this cacophony. But the entire pack of seven adults also began hunting continuously, gorging and carrying home meat in their stomachs to regurgitate. They defended the pups from predators such as grizzly bears. They endured, like so many wolf parents and relatives, the sharp teeth and unrelenting, maddening, playful attacks of their young. Pups grabbing your tail, pups chewing on your ears, pups jumping on your head.

For wolves, a mortality rate of 50 percent in a litter is normal. Out of these twenty-one pups, a triumphant twenty survived. Over the next few years, 21 and 42 stayed together while their large pack eventually split into new packs. We now know that alpha females lead most packs in terms of major decisions, and Rick watched as alpha female 42 "instituted a new era . . . one that was more supportive and respectful" of subordinate members. Rick thought of 21 and 42 as wolf royalty, the "Kennedys of Yellowstone wolves."

In the winter of 2004, the nine-year-old 42 was caught hunting alone and killed by a rival wolf pack. Night after night, the now-silvered 21 howled for her to return. In the spring, he bred with another female and helped feed her litter. But by summer he was dead, high on a ridge where 42's smell lingered in the scent post of a particular tree. Rick McIntyre believed the old wolf had come here still hoping to find his longtime mate.

For the people watching these dramas, the revelation was the individuality of wolves that were compassionate or cruel, generous or selfish, good with kids or not so much, indifferent to humans or frightened to cross a blacktop road, shy, sly, introverted, extroverted. Some wolves died to protect their family. Some wolves left their families behind. These were lives similar to our own in their relationships and affections. At various lectures and public events, Rick has wondered out loud, "Can a wolf in the wild experience what we know as joy and happiness?" The question is rhetorical. He has interpreted or intuited that experience in wolves many times. Similarly, he has seen their sorrow and grief mourning friends and family members.

Research has shown how wolf packs, as a whole, are also singular. Wolves will teach their pups the custom of ignoring domestic livestock and hunting elk, or the other way around, and those pups will teach their pups, who will teach their pups. A pack will use the same denning area for decades, roaming the same valleys and hills. The death of experienced adults in such a pack changes that culture. The death of a parent and the orphaning of yearlings and pups is a major trauma, just as it would be in any family.

Gordon Haber studied Alaskan wolves for forty-three years before he died in a plane crash in Denali National Park. "Wolves are fascinating," he wrote, "as individuals, but what I find unique is the beautiful, interesting, and advanced social structure of an intact group. Fragmentation of a wolf group through [human] hunting disrupts the animal's most prominent characteristic."[4] For Haber, this characteristic was group cooperation, both in parenting and in bringing down large prey.

By 2003, the wolves released in Yellowstone National Park had grown to a population of 174. Wolves had also been reintroduced into a wilderness in Idaho, and wolves continued to disperse from Canada, for an estimated 1,700 animals in the Northern Rockies. Some three thousand wolves lived in Minnesota, Michigan, and Wisconsin. Wolves there were not hunted, although they could be killed in self-defense. Government agents could also kill wolves that attacked livestock or pets. People had learned to live with wolves.

The 1973 Endangered Species Act was an extraordinary step in the conservation of plants and animals. Once protected, being delisted from this federal oversight could put a species at risk. In the states surrounding

Yellowstone National Park, hunters had become accustomed to an over-abundance of elk, and wolves were resented as competitors. In 2005, the management of wolves was transferred from the federal government to state control in Montana and Idaho. In 2008, all Rocky Mountain wolves were delisted from the federal endangered and threatened list. At this point, and based entirely on politics—people for and against wolves—hunting wolves became legal in Wyoming and then not, legal in Montana and Idaho and then not, and finally legal in all three states. From 2011 to 2020, more than 3,500 wolves were shot or trapped in the northern Rockies, many of them Yellowstone wolves crossing the park boundary.

In the fall of 2020, under President Trump, the federal government removed protection of the gray wolf in all the lower forty-eight states. Within a year, under President Biden, that decision was reversed. The Rocky Mountain wolves, however, remained under state management. In 2022, Idaho game laws allowed hunters and trappers to kill up to fifteen wolves each, all year long, using previously restricted methods such as snowmobiles. The state also offered a bounty on wolves. This plan provided for the elimination of 90 percent of Idaho's wolf population. Montana and Wyoming offered their own hunter-friendly version of wolf management. That year, the wolf population in Yellowstone National Park reached a dangerous low as a result of wolves crossing out of the park and being killed.

Every four years, a new president can mean more reversals in federal protection. Then there are the midterms for Senate and House. Finally, there are state elections—who is elected governor. The governor, in turn, chooses the state's wildlife commissioners in charge of setting policy and hiring and firing the state wildlife agency's director. At this level, decisions will be made. Can hunters use night-vision goggles? Can they run animals down with ATVs?

It almost seems wolves should be allowed to vote.

Where I live in New Mexico, the reintroduction of gray wolves has been similarly fraught. By the 1970s, the Mexican gray wolf—a southwestern subspecies—was considered extinct in both the United States and Mexico. In 1998, eleven endangered Mexican gray wolves from zoos were released in Arizona's Apache-Sitgreaves National Forest and Blue Wilderness.

In 2023, an estimated 240 Mexican gray wolves were living in the wild. Wolves fit fine into the ecology—but not into the human culture. Wolves were being shot at and killed. Wolves were being released, recaptured, transported, and sent back to zoos. The wild wolves—now also in the Gila National Forest and the Gila Wilderness—were inbreeding, their genetics questioned. For more than twenty years, the federal recovery program has suffered through mismanagement, bad faith, obstruction from ranchers, and lawsuits by environmentalists.

In 2012, my first wildlife tracking class was a one-day workshop in the ponderosa pines of eastern Arizona. The leader of the workshop, Janice Przybyl, occasionally saw in this forest the tracks of the Mexican gray wolf. Because that's the wonderful thing about tracks: they stay there, even as the wolves move on, hiding from the people shooting at them. Janice was not just a master tracker—she was a superstar of what many environmentalists in North America now call community science. (The well-known term *citizen science* is being abandoned because of its connotations of legal citizenship in a specific country, rather than the original and intended idea of "citizen of the world" or participatory research by the general public. For some environmental justice groups, however, community science still refers to a subset of citizen science that focuses on community-led programs and the health of a particular community). As a graduate student, Janice had explored the use of trained volunteers to look for wildlife tracks, with photos of the tracks confirmed by professionals. This model became the Wildlife Monitoring Program at Sky Island Alliance, a conservation group based in Tucson. The information that Sky Island Alliance gets from its volunteers had already prompted the Arizona Department of Transportation to build two wildlife underpasses and one overpass across highways. Today, agencies and municipalities in Arizona regularly consult with Sky Island Alliance about the needs of such animals as mountain lions and desert tortoises.

At that time, I was researching community science and writing about tiger beetles. For two summers, in collaboration with professional entomologists, I focused on these quarter-inch-long predators with oversize mandibles. Tiger beetles run down their prey, slice, dice, drench the victim in digestive juices, and suck up the ooze with a special strawlike mouthpart. The tiny tiger beetle larvae are equally ferocious, anchoring in their burrows with a special hook on their backs and lunging out at anything tinier than themselves. After dragging the prey into tunnels, the larvae cut them into pieces with larva-size mouthparts. Then they feed, pupate, and emerge as decorative adults, variously patterned and iridescent in the sun. We can still find new species of this insect when we look. Tiger beetles are everywhere.

Years later, I would become a volunteer with Sky Island Alliance. My friend Sonnie was one of the leaders in our volunteer group, working on a transect in the Burro Mountains halfway between her home and mine. We had a list of animals to document: predators such as mountain lions, bobcats, and black bears, whose tracks we often saw, as well as endangered Mexican gray wolves, ornate box turtles, and New Mexico meadow jumping mice, whose tracks we never did. Other volunteers around North America also monitor local species: grizzly bears and wolverines in Washington State and British Columbia, bighorn sheep in Colorado, fishers in Connecticut.

Years later still, I would be emailing Janice during a global pandemic, and she would tell me about her work with the Grand Canyon Wolf Recovery Program. Their mission is to support the natural dispersal of Mexican gray wolves into the thirty-six million acres of remote land that includes the Grand Canyon National Park. They believe wild wolves should be able to move freely into this area without being captured and released south of Interstate 40—the current northern border of where wolves are allowed to live in Arizona. The Grand Canyon Wolf Recovery Program has campouts, film festivals, educational booths, and group wolf howls. They write letters and circulate petitions. It's a struggle. Too many people still seem to hate wolves the way wolves hate coyotes and coyotes hate foxes.

In that first workshop with Janice, in 2012, we were at someone's home and backyard that adjoined the Apache-Sitgreaves National Forest, where a forest fire had burned more than 500,000 acres a year before. Miraculously, the home still stood, surrounded by blackened trunks, a dystopian scene of black spars and poles. A few of the older ponderosa pines were alive and would recover, but many more would not. Although periodic low-intensity fires are necessary for these pine trees, the previous year's fire had burned too hot to be part of that natural cycle. As one fire ecologist said, fire suppression and the buildup of brush and crowded trees, along with clear-cut logging, "sort of broke the structure of the forest." Habitat for some animals like the Mexican spotted owl disappeared over night. Stands of spruce and fir would never return.

I felt depressed about this, and I wanted to feel depressed. But the wildflowers made that difficult. My body kept responding to the profusion of red rocket, blue lupine, white yarrow, purple aster, Indian paintbrush, salsify, daisies, groundsel, and goldenrod. Flowers often come out like this after a fire, in soils briefly rich with the nutrients of burnt trees. Some of these seeds had needed to be cracked open by heat, while other flowering plants were taking advantage of the suddenly opened space. It was like walking through a bouquet. All that scent and sex in the air.

After a lecture and PowerPoint, Janice went ahead of the tracking class into the burnt trees. Her adult learners straggled behind, most of us with gray hair and versions of ADD, chatting about this or that, constantly taking pictures of wildflowers. Earlier in the day, Janice had used plaster of Paris casts to press fake tracks into the ground along the trail and in a nearby streambed. The casts had been made from real tracks by pouring cement into a print and attaching a wooden handle. Our job was to find these fake tracks and identify which animal had made them.

Of course, no animal had made them because they were fake. But we didn't care. My husband and I knelt before such a print in the white sand of the streambed, granite boulders surrounding us, gray and pink rocks knee-high, shoulder-high, speckled like trout, glittering with quartz. Fool's gold shone at the bottom of the water. Mica sparkled in the sunlight. As Peter and I studied a fake print, we felt the excitement of any treasure hunter. Children on an Easter egg hunt. Me, me, I found it!

And then the joy of pattern recognition, matching something recently learned to something on the ground. This was an oval track slightly less than two inches long, H in the negative space, small palm pad, dainty toes pointing upward. Gray fox. We crouched. We conferred. We moved up the streambed. Here was another track. Five toes, fused palm pad and heel of hind foot looking like a human footprint, the smallest toe on the wrong side. Bear. Me, me. I found it. And another track, so obviously feline. The square block into the square hole. The round block into the round hole. We scrambled from print to print. "Over here!" "Here's another!" Competence. Discovery.

It's easy to make fun. Honestly, though, I'm pleased today with just an echo of that.

To be nice, Janice had also faked the track of a Mexican wolf.

By now, the print of a gray wolf will seem familiar. Those four distinct toes, that canine symmetry—one half looking like the other half—with front feet slightly bigger and rounder, an X or H in the negative space, and the claws showing reliably. The claws of a gray wolf tend to be large, similar to those of domestic dogs. The top of the palm pad has one lobe or point and the bottom has two. Most obviously, the tracks of a gray wolf are big, measuring more than three inches to almost six inches long and almost three inches to five inches wide. The tracks of a Mexican gray wolf are somewhat smaller.

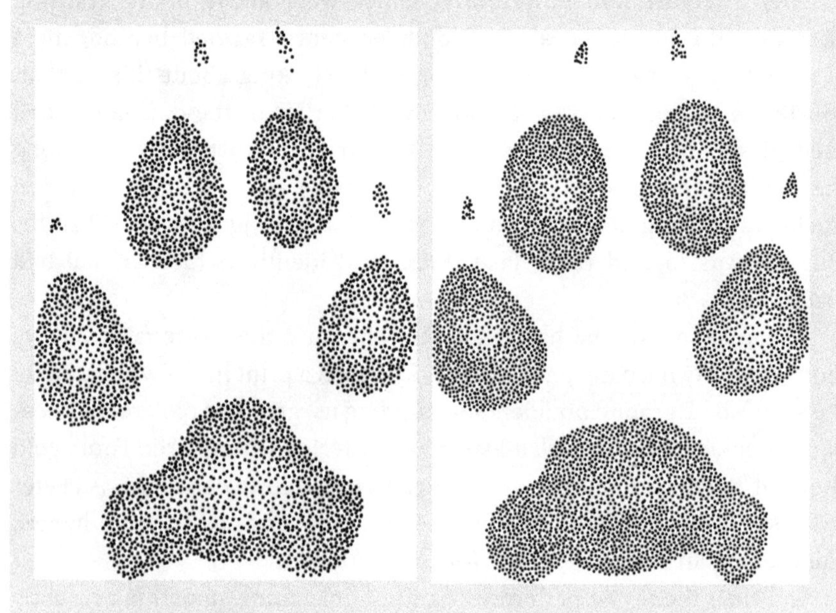

FIGURE 5.2 Gray wolf, front and hind

WILDLIFE TRACK AND SIGN CONVERSATION #208

"So this track is about a half inch long, with five toes not connected to the palm pad, toe 1 pointed to the side, toes 2, 3, and 4 pointing up, and toe 5 to the side. I think it's a harvest mouse."

"Harvest mice have slightly bulbous toes. I don't think these are bulbous."

"*Bulbous* is becoming one of my favorite words."

"The toes are pretty thin."

"The toes of the endangered New Mexico meadow jumping mouse are thin. That's on our list."

"That *is* on our list."

Silence.

"Long-tailed voles also have thin toes."

"Voles are more about grassland."

"Meadow jumping mice are more about wetlands."

"Jumping mice. Saltatorial. Adapted for leaping."

"It's too dry here for the meadow jumping mouse."

"Not vole. Not harvest mouse."

"Pocket mouse?"

"Deer mouse?"

"Northern pygmy mouse?"

"I don't know what it is."

"Me neither."

"A mouse of some sort."

6

We'll Lift That Lion Right Off That Limb

If you ever feel threatened by a mountain lion, also known as a cougar or puma, the best thing to do is threaten back. Act big, stand tall, get mad, say "No." Mountain lions will usually demur. If the animal doesn't, if you are attacked, fight.

In 2019, a thirty-one-year-old jogger in Colorado heard a rustling sound, turned, and was "pretty bummed" to see a mountain lion chasing him. He stopped, waved his hands, and screamed "like a barbarian," but the mountain lion continued straight for his face. They fell, rolled down a hill, wrestled, and the jogger managed to choke the animal to death. Initially, the news report was that brief and generated a certain fantasy for those of us who jog or hike in landscapes that include mountain lions. It's not that we thought we would be attacked. It's not that we wanted to choke an animal to death. It's just that each of us has had a moment—hearing a sound, turning, wondering, seeing nothing. If we did see something, we knew we should project, "I'm too much trouble to eat!" We should raise our arms, shout, struggle, wrestle, and—now Plan B—go for the throat.

Soon we learned that this mountain lion was a starving cub, maybe five months old, orphaned, and weighing some forty pounds. If the mother had been alive, she would have been bringing food to this male and his two siblings, who were eventually captured and sent to a rehabilitation facility. She would have been modeling how to hunt for at least another year, showing the inexperienced juveniles how to be quiet, stay quiet, move quietly. She would not have attacked a man jogging rather than her usual prey of mule deer. She would not have used a frontal attack. Also, weighing a muscular hundred pounds, she would have been too powerful for a 150-pound man to hold down as he pressed on her throat with his foot. He would have been less likely to survive this attack—although most people do survive a mountain lion attack. That's much more common than not.

I feel I should start again. Mountain lions rarely attack, injure, or kill people. Statistically, you have a better chance of dying from a champagne cork, two dozen deaths a year, usually at weddings, or a golf ball, hundreds of deaths a year, usually near golf courses. Humans are not what mountain lions like to hunt and eat. Humans make mountain lions want to go hide. Mountain lions back down from yells and hand waving because they are, in fact, timid and cautious predators, having evolved with other predators much bigger than themselves. At the end of the Pleistocene, mountain lions (80–180 pounds) coexisted with saber-toothed cats (350–600 pounds), American lions (400–550 pounds), scimitar cats (350–400 pounds), American cheetahs (150–200 pounds), and jaguars (120–210 pounds), not to mention the terrifying great bear (600–2,000 pounds) and dire wolves and gray wolves. Mountain lions learned to be stealthy not just with prey but as prey.

When the other big cats went extinct, mountain lions did too—except for a small population in South America that eventually and enthusiastically recolonized the rest of the continent. Thousands of years later, after Europeans reached the New World, they killed mountain lions as ruthlessly as they killed coyotes, wolves, and other predators. In the eastern and midwestern United States, these animals were extirpated (the local extinction of a species). Today dispersing adults and juveniles are beginning to reenter their former range.

The western United States still has an estimated thirty thousand mountain lions. About a hundred endangered Florida panthers, a subspecies, may live in the swamplands of Florida. Canada has some four thousand mountain lions, almost all of them in British Columbia. No one knows how many mountain lions live in Mexico and Central and South America—perhaps another fifteen thousand. Because of that uncertainty, the IUCN lists this species as of least concern, with a decreasing population trend.

I have seen a wild mountain lion only once, glancing up from the desk in the office where I write as the animal crossed from the bird feeder to disappear around the side of the barn. Most of my friends also have a story:

two mountain lions in their headlights, a mountain lion posed on a ridge, a mountain lion watching from a bush. My husband once watched a large female slowly cross the irrigation bridge thirty yards from our house. Her head turned toward him, seconds passed, and he swears he saw the calculations click behind her eyes: prey, distance, cost, benefit. We know this lion was female because a day later her body came floating down the irrigation ditch, shot from a neighbor's tree into the water. That was sad but not surprising. She had been eating goats, jumping over a tall fence and jumping back, the goat gripped by the neck. She had become too visible.

Of course, I see mountain lion tracks. Mountain lions use nearby trails, dirt roads, and the corridor of the Gila River to travel through their territory or to new territory, since juveniles must eventually leave their mother and find someplace else to live. I often saw their prints in the transect I monitored with Sonnie for Sky Island Alliance. I could go outside my house right now and, with time and effort, probably find a track within a mile. If you live in the American West and know where to look, you will see their sign, too. Even if you live in Los Angeles or Tucson or Denver or Seattle.

Like canine tracks, all feline tracks have certain distinct qualities.

They usually register four toes. The front feet have a fifth toe higher up on the leg that registers in deep substrates.

They are asymmetrical. One half does not look like the other half.

Of the four teardrop-shaped toes, one middle toe often extends slightly past the other one. Which toe is the "leading toe" tells you if the track is from a right or left foot. Think of the leading toe as equivalent to your middle finger. Put your right and left hand above the track and see which one, minus thumb, matches.

The claw marks are usually absent, unless the animal is using them for traction in a chase or in mud or for some other reason. Claws are on a higher plane than the base of the toe pad and can look like pinpricks (smaller cats) or like the point of a knife slicing into the dirt (bigger cats).

The negative space usually has the shape of a C pointed down. This is easy: C for cat. If you tried to draw an X in that space, you would bisect the palm pad.

The palm pad of a feline is relatively big. You can fit jigsaw-style most of the four toes into its space. Compare this to the front track of a wild canine, with a palm pad into which you can usually fit only one to two and a half toes.

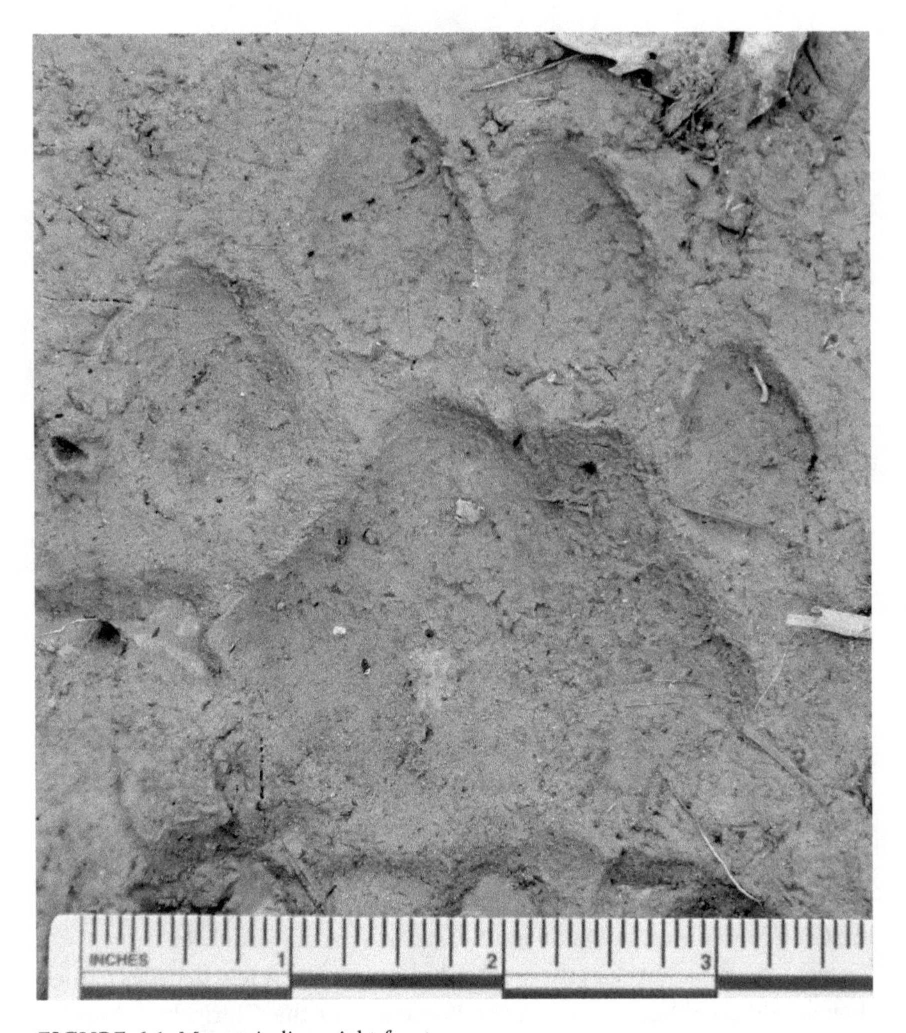

FIGURE 6.1 Mountain lion, right front

The feline trapezoidal palm pad has two lobes at the top, not one. Sometimes these two lobes register as a flat line. The three distinct lobes at the bottom often register.

In the first half of the twentieth century, three other cats—jaguars, ocelots, and jaguarundis—also lived in southeastern and southwestern United States. The last female jaguar was probably shot in Arizona in 1963, although traveling males are sometimes seen today. The last documented

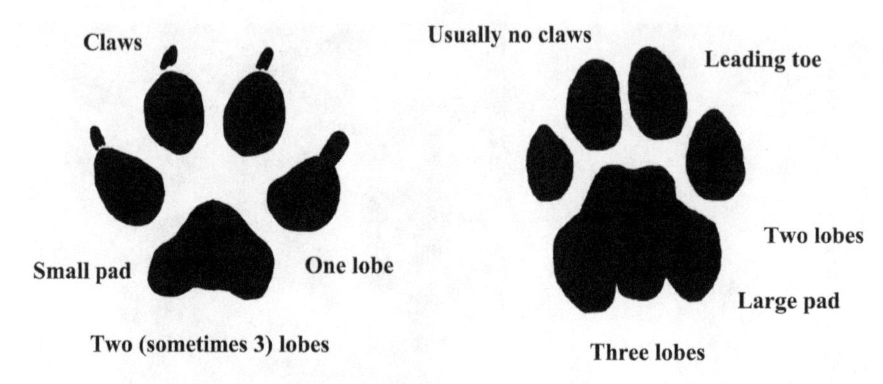

FIGURE 6.2 Domestic dog and mountain lion

jaguarundi, a brown weasel-looking feline twice the size of a housecat with short legs and long neck, was killed by a car in 1986. A population of fewer than sixty ocelots is believed or hoped to be in southern Texas. All three cats still live in Central and South America, with ocelot and jaguarundi listed by the IUCN as of least concern, with a decreasing population trend. The jaguar is listed as near threatened, a step above least concern and a step below vulnerable.

These animals became extinct in the American Southwest partly because of men like my great-uncle by marriage, Ernest Lee, and his brothers, Clell, Dale, and Vincent, as well as Ernest's father-in-law, my great-grandfather, and various aunts and cousins who lived in the small town of Paradise in the Chiricahua Mountains of southeastern Arizona. The Lees came from Texas to Arizona in the early twentieth century. Just a few decades earlier, they would have been entering the homeland of the Chiricahua Apache, who had been largely conquered and shipped to Oklahoma. My great-grandparents also came from Texas to Paradise, where they failed at running saloons. Meanwhile, the Lees were failing at farming. When John Lee died in 1916, his wife and seven children opened a stage line hotel. Eventually the handsome oldest boy married my great-aunt Bertha, twelve years younger than himself. They were dirt poor, but with horses, dogs, and guns.

The Lee brothers killed more than a thousand mountain lions, a thousand bears, and 124 jaguars, as well as uncounted bobcats, ocelots, and jaguarundis. No one counted the coyotes, either, although in one photo Ernest stands next to forty of them nailed to a barn. Mountain lions are most easily killed with a pack of dogs, which chase the cats into a tree from

FIGURE 6.3 Great-grandfather John Bendele

which they can be shot down. Necessarily, the Lee brothers would become as admired for their hounds, developing and selling the bluetick breed, as for their ability to hunt. At that time, the two were one and the same.

In a coauthored memoir, Dale Lee remembers being part of a mountain lion hunt when he was thirteen. While his brothers' dogs bawled beneath a big tree in a small canyon, Dale climbed the nearest bluff for a better view. "And boy, I thought that was the most beautiful sight I'd ever seen. I'd got up on the hillside and was looking down on the lion, and the rest of them were down at the tree looking up."

I can imagine myself with that shirttail relative, a hundred years ago. The Chiricahua Mountains are now a national monument and national forest, and the scenery today is much the same: blue hills layered to the horizon, bluffs the color of dried blood, plants like tiny knives—needle-tipped cholla and barrel cactus with fishhook spines. The big tree in this canyon is a sycamore, with the white sensual limbs of a goddess. There's the smell of chalky dust and maybe something herbal and tangy. In the tree, a lion crouches, frightened, defiant, built like all cats to ambush and kill in a burst of speed, with a long tail for balance and a compact head for biting strength.

Dale continues his story:

"After a while, Ernest looked around. 'Dale, where are you?'

"And I said, 'Here I am. I'm up here.'

"'What're ya doin' up there?'

"'Well,' I said, 'as long as I stay here, I don't believe that lion can get aholt of me. So I'm up here on the hill.'

"And, aw, they really laughed. They got a big kick out of that. But . . ."

Dale repeats himself now. He can't help how he feels.

" . . . but it was really a beautiful sight. That was the first lion I ever saw caught by the hounds."[1]

In 1926, Dale graduated from high school and joined his brother Clell working for the State of New Mexico killing predators. Clell was paid $125 a month and his helper, Dale, $30. The teenagers split the money and worked as a team. Often they were called to ranches to kill mountain lions that might or might not be eating cows. Although this was a lot of fun, brother Ernest would soon have a family to support, and he needed a somewhat better income. Eventually all the Lee brothers began guiding trophy hunters, with Ernest the manager and brains of the business. His mother and his wife, my great-aunt, were in charge of feeding the dogs, cooking huge pans of cornbread on a wood-burning stove. With some empathy, Dale remembered, "After a few years, we had forty or fifty hounds, and when two women had to cook cornbread for them every day, as well as feed a big family, well, that was a chore."[2]

During World War II, Dale guided officers back and forth to the front lines for the invasion of Normandy. By the time he and his brothers returned to professional hunting in the 1940s, many western states offered a bounty for mountain lions that ranged from $50 to $100 per head. The ambition was to get rid of these predators, much as we had gotten rid of wolves. Bounties had been successful with wolves. So had poison. So had steel leghold traps.

FIGURE 6.4 Clell Lee, great-uncle Fred Bendele, and Dale Lee

Somehow, though, the big cats survived.

The Lee brothers continued to raise dogs and take out sport hunters. One photo shows Dale and a hound beside five mountain lions in Oregon. A Christmas card has Dale and a hound beside seven jaguars in Venezuela. By the 1950s, the Lee brothers had begun to specialize in these spotted jaguars and ocelots, first in Mexico and then Central and South America.

Dale's stories always included the injuries to the dogs, which were many and horrific. Sometimes he joshed about the jaw-dropping stupidity of his clients, as well as the pranking between him and his brother or between him and the cranky old cook, my great-grandfather. The stories usually ended the way his first hunt had ended, "The jaguar was treed good and high, and it was a beautiful animal up in that tree, a good, big male. . . . 'Okay, boys,' I said, 'we want that thing dead, so I'm gonna say one, two, three, and you guys shoot him from that side, and I'll shoot him from this'n, and we'll lift him right off that limb.' "[3]

At this point—before this point—I can no longer imagine myself with these distant relatives. I still admire parts of Dale's life. He lived outdoors as though outdoors was home, stoic in the face of physical discomfort and in the acceptance of nature's violence and death. He was a man who knew things: where black bears denned, where his dogs liked to be scratched, where he could find pinon nuts for the women to shell. Particularly Dale knew mountain lions, their speed and gait, what they ate, how they digested their food. He knew that mothers brought their kittens both dead and live prey and took them to kill sites, teaching them to hunt. He knew that mothers, daughters, and sons traveled together for as long as two years and that a resident male lion might defend his territory by killing transient juvenile lions. Although most lions died young for a variety of reasons, Dale had known some to live more than a dozen years until their teeth and sinews finally gave out. He knew that mountain lions mostly stalked and ambushed deer, but they also ate elk, cows, coyotes, raccoons, rabbits, and mice.

In 1965, when he was fifty-seven years old, Dale Lee began working as a game officer at Fort Huachuca in southern Arizona. Into his seventies, he continued to hunt—seven bears, he once boasted, in the first two weeks of bear season. He also consulted with wildlife biologists working to protect mountain lions and their habitat. One of those biologists, Harley Shaw, became a friend of mine and told me that Dale gave him a lot of good advice and useful information. "Truth is," Harley emailed me in 2020, "I've always hated to see the wonderful cats killed, but I can't help admiring the skills developed by the old dryland hunters."

Harley believes that houndsmen like Dale offered important resistance against the widespread use of poison, fearing the consequences for their dogs. Regarding leghold traps, Dale said he had too much respect for the mountain lion to "hang steel on its foot." Almost all lion and bear hunters stopped supporting the extinction of predators when they realized the obvious: no more lions and bears meant no more lion and bear hunting. Later, with the passage of the Endangered Species Act in 1973, federal programs could no longer support the eradication of mountain lions—or any animal.

At the state level, as usual, the protection of mountain lions varies. As of 2023, California has continued its ban on hunting mountain lions for sport. In Colorado, the season is five months with a limit of one mountain lion per hunter. In Texas, mountain lions can be killed year-round, in any number and at any age, including spotted kittens. In Canada, the province of British Columbia has a seven-month season, hunting dogs are allowed, and each hunter can kill two lions. The province of Alberta has two seasons, one that allows dogs and one that doesn't.

The rules are real, and almost every hunter takes them seriously. Dale Lee took them seriously. Unknowingly he was part of a movement and tradition called the North American Model of Wildlife Conservation, or NAM.

The articulation of NAM is fairly recent. In the mid-1990s, hunters continued to decline in number even as wildlife advocacy groups were increasing. The Canadian biologist and hunter Valerius Geist joined with others to promote a narrative in which recreational hunters in Canada and the United States had rescued wildlife from extinction and still served as wildlife's best protectors. This somewhat revisionist history starts in the late nineteenth century with President Teddy Roosevelt. Seeing the destruction caused by unregulated hunting and land development, Roosevelt began a system of public lands, laws, and funding that future generations of hunters and self-described conservationists built on. Canada passed similar legislation protecting its wildlife.

The focus on both consumptive use and regulation was further developed into seven "pillars": (1) Wildlife is a public resource and held in public trust. (2) Markets to buy and sell meat and parts of game are illegal, although a market for the skin of furbearers remains. (3) Access to wildlife for hunting is regulated through set hunting seasons, bag limits, license requirements, etc. (4) Wildlife can be killed only for a legitimate purpose. (5) Wildlife species are an international resource. (6) Science is the tool for determining wildlife policy. (7) Hunting and fishing are a democratic activity open to all.

NAM is widely celebrated by government agencies and hunting groups. The U.S. Fish and Wildlife Service says, "American sportsmen and sportswomen are the backbone of the North American Model of Wildlife Conservation that is admired around the world." The Association of Fish and Wildlife Agencies echoes, "The North American Model of Wildlife Conservation is the world's most successful system of policies and laws to restore and safeguard fish and wildlife and their habitats through sound science and active management." Hunting groups and conservation groups linked to hunting and fishing are even more enthusiastic.

Although the history presented by NAM contains many truths, critics complain that it almost completely leaves out the contributions of environmental groups, as well as individual women and people of color. NAM proponents also tend to overstate some of their recovery successes. They ignore the historic attempts to eradicate predators and the continued hostility toward predators. And they ignore the decline of nongame or nonhunted species such as bats, the prairie dog, and pollinating insects.[4]

In the United States, the North American Model of Wildlife Conservation means that state wildlife agencies are strongly motivated to please the hunters, trappers, fishers, farmers, and ranchers who oversee and help pay for their budgets. In many states, for example, the limit or quota of mountain lions to be killed is partly based on any perceived threat to deer and elk populations or to livestock. In the language of wildlife management, mountain lions are harvested like a crop. In the 2022 season, in my state of New Mexico, 164 male and 55 female mountain lions were sport harvest (trophy hunting), 15 were depredation kill (mountain lions suspected of preying on livestock), 17 were killed for bighorn sheep protection (mountain lions in areas with reintroduced bighorn sheep), and 11 were killed in accidents with vehicles. The total of 262 dead was about 10 percent of an estimated population of 2,500 mountain lions.

Harley Shaw complains, "There still isn't an accurate method for monitoring cougar numbers, yet recommendations for 'harvest levels' are often expressed in terms of a percentage of an unknown population."[5] A recent analysis by Arizona's wildlife agency used a complex model, with field data from many sources and across different periods, to come up with the same number that Harley had suggested forty years ago. That was a pretty crude guess, he admitted then, based on two study areas. He thinks the guesses for population are still pretty crude.

An agency that regulates hunting and trapping would tend to overestimate wildlife numbers, implying that we have plenty of animals to hunt.

Adding up the estimates of government agencies in the United States comes to a hopeful high of forty thousand mountain lions. A conservation group interested in protecting a species might underestimate, suggesting that we don't have enough animals to carelessly kill. The earlier estimate of thirty thousand mountain lions in the United States comes from the Mountain Lion Foundation, whose mission is "to ensure that America's lion survives and flourishes in the wild." The Humane Society has chosen a different approach, looking instead at the availability of mountain lion habitat. Right now, again in the United States, areas with breeding populations could support forty-three thousand mountain lions. The Humane Society states that trophy hunting is the main reason we don't have these missing lions.

The question then: Are thirty thousand or forty thousand mountain lions too few, too many, or just right?

Another question: should we kill mountain lions as recreation, hunting them with dogs, shooting them from trees, and displaying them as trophies?

The argument for trophy hunting is that the money from this sport helps fund wildlife agencies and conservation programs, as well as providing jobs for hunting guides. The hunters enjoy themselves, and many experience a profound satisfaction and connection to nature. Hunting big charismatic predators is a legacy and primal desire. In the diversity of human behavior, there is room for this one. Trophy hunting might limit the overall mountain lion population in North America and affect local populations, but the number of lions seems stable and the species is not endangered. Importantly, although we admire and respect mountain lions, they have no intrinsic value except their value to us.

The argument against trophy hunting is that this last sentence is not true. Like wolves, mountain lions have intrinsic value to themselves. They have individual lives we should not savage. Beyond that individual life, trophy hunting alters the social structure of mountain lions. When a resident male dies, the transient males entering his territory may kill cubs not their own, hoping the mother will breed again. Sometimes they kill the mother who is protecting her cubs. When a female is killed by a trophy hunter, any young she is training suffers and possibly starves. Trophy hunting can affect people, too. Older and more experienced mountain lions are the animals least likely to become a problem in terms of conflict with humans and their livestock. Orphaned cubs and yearlings are more apt to choose inappropriate prey, like the jogger at the beginning of this chapter.

People on both sides can agree that mountain lions have ecological value. In his book *The Cougar Conundrum*, Mark Elbroch calls them "ecosystem engineers," or species that modify or even create habitat for other animals. Like wolves, mountain lions provide large carcasses that feed other mammals, birds, and insects. They control populations of prey. A return of mountain lions to the eastern United States would help reduce their swollen herds of white-tailed deer, culling sick deer (especially those with chronic wasting disease, which can spread to other ungulates), resulting in healthier, more sustainable deer herds. Again, people would benefit because deer actually kill and injure more people—in the form of car accidents—than any other wild animal.

One summer morning, I was running along a dirt road on Sacaton Mesa, a broad elevated expanse of land near my house. The view here is 360 degrees, rolling hills ahead and to the left, the Mogollon Mountains on my right. Sacaton is a large perennial bunchgrass, which overgrazing on this mesa has replaced with mesquite and snakeweed, punctuated by yucca. In the distance, desultory patches of other grasses were still being grazed by cows. Suddenly, three dogs came rushing toward me. No, four! Five! I was surrounded! These weren't Michelle's German shepherds, and they weren't the mix of heeler and pit bull that bark at me when I am bicycling. These were, oh my gosh, I had never seen these dogs before. But their tails were wagging. And they had that goofy look. They were hound dogs.

A man I had also never met before walked toward me, calling the dogs, and I was saying, "Oh, it's fine. They're fine!" The man pushed something, and as we got closer, I saw it was a baby stroller. Inside, facing me, sat a solemn boy of about two. The man was also young, maybe in his thirties, my height but with a big cowboy hat. I was delighted. I had been running on this road for many years and had never seen such a sight. We smiled at each other as I jogged past. By now the dogs had found something else to interest them.

And then I stopped, turned, and spoke. He stopped, replied, and told me he was leasing this land for grazing. We exchanged some code words. "You're leasing from the Shellys?" "Oh, from the Agnews." These were the names of ranchers who lived in the Gila Valley, proof that we both belonged here. I asked—since he had hound dogs, maybe bluetick dogs—if he had ever heard of the Lee brothers. And, of course, he had. Casually I mentioned that Ernest Lee had been my great-uncle by marriage. The young man's face lit up. It was like meeting a celebrity. Here on Sacaton Mesa.

I wondered what he hunted with his dogs. He said he used to hunt bear but now he hunted mountain lions. Bears were too much trouble. I nodded. If you had a pack of hound dogs in southwestern New Mexico, you probably hunted mountain lions. Perhaps he hunted mountain lions that were killing cattle or that had threatened pets or people. More likely, though, he guided trophy hunters. We said our goodbyes. I would see him a few more times, in his truck, once with the stroller. We waved. Then I stopped seeing him. Maybe the lease didn't work out. Maybe our paths just didn't cross. I felt relieved and guilty. If I had seen him, I knew I wouldn't ask what I wanted to ask.

More recently, I almost had this discussion with a sixth-grader. I was in his classroom as a guest, teaching the basics of wildlife tracking. We had just seen skunk and raccoon tracks close to this rural school, but I was disappointed not to find any sign of coati, a relative of the raccoon ranging north of Mexico to Arizona, New Mexico, and Texas. These omnivores weigh about ten to twenty-five pounds, with long white noses and long banded tails, masked faces, and thick coats of brown or red or gold. They live in matriarchal groups, bossy moms and aunties and a dominant female leader, everyone always talking to each other, chittering, churring, murmuring, barking. Such groups have a kind of magical realism.

I was showing the class a photo of a bachelor male, a charming, black-eyed, masked bandito. Bachelor coati travel alone most of the year, kicked out by that matriarch or maybe leaving in a huff or maybe ambling off in genial fashion. The sixth-grader raised his hand with a question. He had a blond crewcut, hazel eyes, and a narrow face. He wore a checkered cowboy shirt. "Can you kill them?' he asked. At first I wasn't sure what he meant. Then I filled in the missing words. Can you kill them legally? If so, when and how many? He wanted to know the rules.

I explained that coatis—like pine martens, river otters, and black-footed ferrets—were protected furbearers in New Mexico and could not be killed. In Arizona, however, you could kill one a year, during their designated season. The boy nodded, more to himself than me. He was a serious child and polite.

In both cases, the young man, the young boy, I wished we could have talked more.

Mountain lion tracks usually register four toes. One "leading toe" or middle toe is often slightly above the other middle toe. The tracks are asymmetrical: one half does not look like the other half. Claws are not usually present. The length of a track ranges from slightly over two and a half to slightly over four inches long. The four teardrop-shaped toes form a half circle above the palm pad. The negative space between the toes and palm pad is an upside-down C or sometimes an H in the hind track. An X drawn in that negative space would bisect the toes or palm pad. The palm pads are relatively large, sometimes holding all or most of the toes, with two lobes at the top of the pad and three lobes at the bottom. Front tracks are rounder and larger than hinds and more asymmetrical.

A heavily clawed Toe 1 is reduced and raised up on the front legs. Used in climbing and holding onto prey, Toe 1 may register in deep substrates. A heel or carpal pad is also higher on the front legs, also used in gripping prey, and may register when the animal is moving fast or in deep substrates.

There are a number of subtle differences between the tracks of an adult male and an adult female mountain lion. Elbroch and other trackers use one easy criterion: if the palm pad of the hind track is more than two inches wide, the track is probably that of a male.

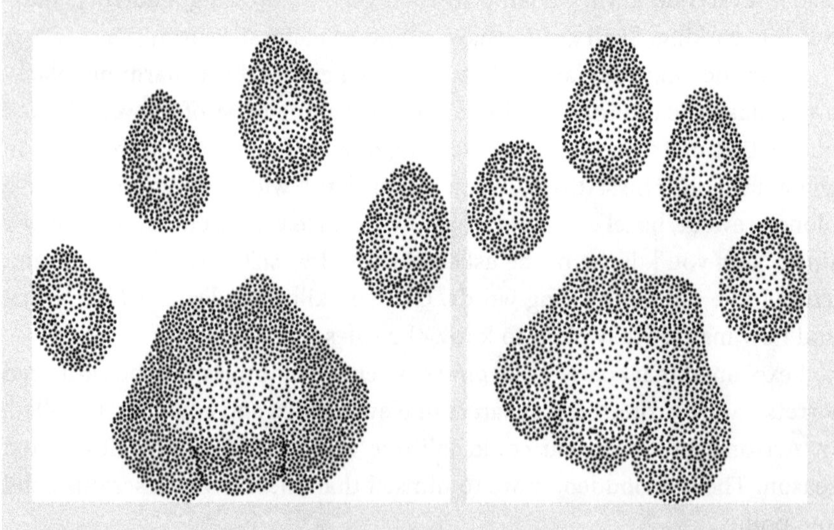

FIGURE 6.5 Mountain lion, left front and right hind

WILDLIFE TRACK AND SIGN CONVERSATION #162

"Look at the width of that pad."

"Maybe a male."

"He's walking."

"This print is rounder, larger. Maybe a front foot."

"Right front foot and right hind foot."

"An overstep walk. Walking fast. Hind foot landing ahead of front foot."

"A juvenile dispersing?"

"Cruising down the river."

"Ole Man River."

"He just keeps rollin."

"Moon River."

"Wider than a mile."

"Take Me to the River."

"Drop me in the water."

Various noises.

"What are you doing?"

"I'm getting closer to the ground."

"What do you see?"

"I can *really* see the three lobes at the bottom of the pad! The edges seem crisp, not much erosion."

"Cats also have sweat glands on their paws."

"Yes! I can smell . . ."

"What?"

"Nothing. Joking. I smell dirt."

7

Bobcat and Lynx

If you have watched house cats, you have seen bobcats. Obviously, they are not the same species. Adult bobcats are generally larger, with males weighing fifteen to forty pounds, females somewhat smaller, coats spotted, six-inch tails, pointed ears with black tufts, and hair in a whiskery ruff around the cheeks. But bobcat kittens, with their round blue eyes, have the same manic innocence as domestic kittens. Young bobcats play together like your cats, ballerina movements, chest and paws raised in slow motion—and then the race begins. Next time you look, they're grooming each other. Tongues licking into ears, a kinetic hypnosis. Bobcats also purr with that grumbling seduction, drilling into your maternal instinct, filing down any species mistrust. (Mountain lions purr, as do cheetahs. But not African lions, tigers, or jaguars.) Like your cat, a sleeping bobcat becomes the Buddha. As with many house cats, too, the eyes of adult bobcats are no longer blue but glow-in-the-dark yellow.

Ancestors of the bobcat crossed the Bering land bridge into North America at least two million years ago. A second group of these ancestors settled in the north and developed into the Canada lynx. Bobcats and lynx can hybridize but mostly do not, the lynx being suited to cold and specializing in snowshoe hares, the bobcat preferring a warmer climate and a variety of prey. Historically, the bobcat's range extended from southern Canada to central Mexico, from the West Coast to Maine and down to Florida. That's

FIGURE 7.1 Bobcat
(photo by Elroy Limmer)

where they live today, too, in forests and deserts and swampland. They eat white-tailed deer. They eat kangaroo rats. In cities and suburbs, they eat ground squirrels and pocket gophers. Based on a 2010 study, the IUCN estimates some 2–3.5 million bobcats in the United States alone.

When you examine a bobcat track—or any other—consider lying down on the ground to check the dirt edges made by the toes and palm pad. Are they crisp and clear or soft and rounded by weather and gravity? For comparison, you might put your hand in the same dirt or sand, also known as substrate, and see what those edges look like. Maybe you have prepared beforehand an area of this soil, made marks, and watched those marks degrade over hours and days. In that case, you know better the effects of rain, wind, time, weight, falling particles. If you are like me, however, you are not that obsessive. You only think once in a while that this would be a good thing to do.

Morning and early evening can be the best times to see this fine detail, when the sun is at a low angle. Morning is certainly the best time to see last night's tracks at their freshest and to imagine that night shift when a kit fox feels more protected and the bobcat is out hunting.

Imagination is key to tracking. Paul Rezendes, author of *Tracking and the Art of Seeing*, begins, "Tracking an animal is opening the door to the life of that animal."[1] Mark Elbroch is credited as advancing the field of tracking through his meticulously researched field guides, which combine experiential learning and technical detail. Mark also says, "The competent tracker is both scientist and storyteller. You must critically observe, collect good data, and avoid rash conclusions, as well as use your imagination to interpret and celebrate the signs you've discovered."[2] Louis Liebenberg, founder of the internationally renowned CyberTracker program that evaluates the skills of trackers, believes that wildlife tracking is the origin of science, with early hunters and gatherers naturally using scientific principles. But even Louis writes, "Tracking is not strictly empirical."[3]

My first tracking mentor, Janice Przybyl, agrees that you use what you know about an animal and weave that into a story of the signs left behind: size and appearance of tracks, gait and speed indicated by tracks, bones and fur in scat, scrapes over feces and urine, bent branches, depressed grass. Sometimes you see where an animal sat or lay down. At that point, Janice says, sit down with them. Imagine not what they are thinking but what they are feeling—hungry or sleepy. Not what they are planning to do but what they will do—groom themselves or digest their food. Don't try too hard. Look around instead.

A bobcat will spend considerable time sitting down, looking around, rotating position, changing the view, creating a hunting bed. Sometimes the leap toward another animal will come straight from that bed. Sometimes instead you see the prints of a stalking bobcat and imagine the crouch, limbs so lowered as to look dislocated, head stretched out, whiskers spread, ears pointed forward—the body simultaneously tense and liquid with premotion. The actual motion that follows is incremental and almost painful to watch. Then, at a certain distance, you see scuffed dirt. The rabbit seized, pinned, bitten at the back of the neck. Like other felines, bobcats have teeth that fit exactly into the spaces of their prey's spinal cord.

Today I have found a series of tracks that show a bobcat moving toward the brushy Gila River. This is a typical slow and quiet understep walk, hind foot falling behind the front foot. This is how bobcats travel while hunting. For a bobcat, the tracks are relatively small and oval, the pad relatively small, the toes relatively dainty, with a relatively large negative space between pad and toes. This might indicate a female bobcat. It is the right time of year and place, so I imagine a female with kittens. That would be likely if she were

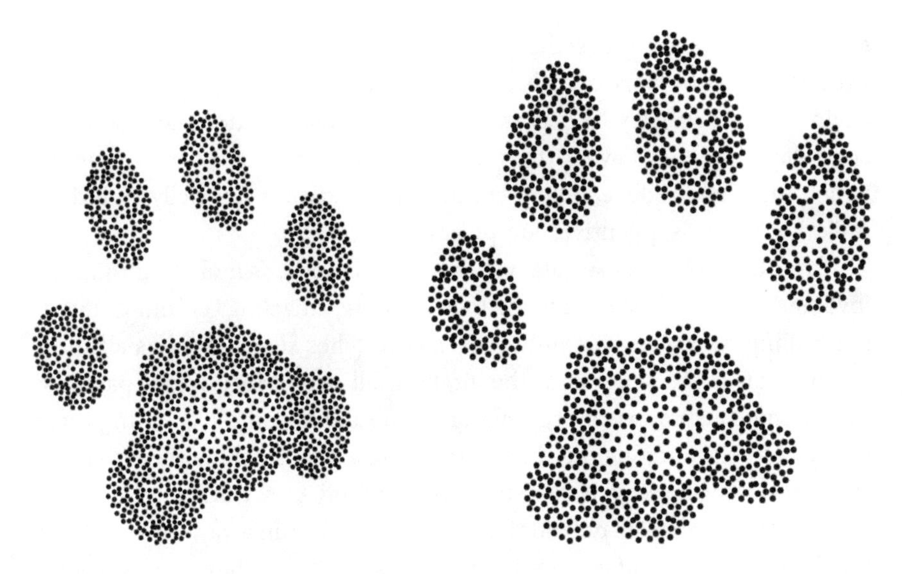

FIGURE 7.2 Bobcat, left front and left hind

two years or older. If I were to see smaller prints here by the adult print, my imagining would be confirmed. I don't. But I let myself enjoy the idea of a mother with kittens still younger than three months in their den. She's looking for a rabbit to take home. Or a mouse. They'll have fun with that.

Like other members of the cat family, the bobcat is a dedicated single parent, staying alone in the den the first days after birth, nursing and licking, keeping the den clean by eating the placenta, feces, and any stillborns. Litters average three kittens. Within a few weeks, these kittens open their eyes and start engaging with the world, like pop-up restaurants. They are nursed for two months and left alone to play while their mother goes hunting. After three months, if they are still alive—if they haven't wandered—they start to accompany her. The white underside of her black-tipped tail and white patches on the backs of her ears flick ahead, visual signs for them to follow. At about nine months, the young bobcats start hunting on their own. Sometimes the juveniles leave suddenly, decisively, to find a new territory. Sometimes they linger. Occasionally, they stay with their mother into a second year.

Bobcats and mountain lions are often described as solitary animals. Bobcats are "solitary," says the seminal textbook *Urban Wildlife Management*. Wikipedia dutifully repeats, "Like most felines, the bobcat is largely solitary." But really that's the patriarchal perspective, taking the male's experience as the norm. With her juveniles dispersed, the female breeds again—from February to September here in New Mexico—and soon, in sixty days, she will have kittens again. Each claims an exclusive nipple, which may help reduce sibling conflict. Each kneads her belly to induce milk. Each mewls, plaintive, compulsory.

If you have seen house cats, you know this mother's look of unthinking love and all the constant acts of love: feeding, nuzzling, cleaning, nudging, calling, soothing. Within weeks, she watches sleepy-eyed as all three tumble out of the den, bat at the air, and fall over. More weeks pass and there's the constant work of bringing them rabbits and mice, letting them play with their food, shepherding them along a trail, teaching them to hunt. Vigilance, safeguarding. If there isn't enough food, they will starve. If a male bobcat attacks, she will defend them. Or she might run away. These are not the decisions of a solitary life. If her kittens don't die, they won't disperse for months. Soon after, the cycle begins again. Solitary? I don't think so.

Bobcats are not threatened or declining as a species. But they are a harvestable crop, susceptible to what that means. The market in wild fur goes up and down as fur goes in and out of fashion and as ranch fur (mink and other animals raised in cages) replaces wild fur. In the twentieth century, when international treaties began to protect endangered cats such as jaguars, ocelots, and cheetahs, furriers turned to bobcat and lynx, especially prizing the bobcat's white black-spotted belly fur. Between 1998 and 2002, North America exported some 119,000 bobcat skins. In 2013 alone, the United States exported 65,000. A 2020 fur report noted that a high-quality bobcat pelt could bring in $300–400, with the furs mostly worn by affluent women in Russia and China. Then, during the COVID-19 pandemic, farm minks were killed because they could transmit the coronavirus, and the price of a bobcat went up again.

Those bobcats were mostly caught in leghold traps, the force of two opposing steel jaws slamming together on a leg or paw. More than a hundred countries, including Russia and China, have banned leghold traps as inhumane. They are legal, however, in Canada and much of the United States. In leghold traps, the injuries include torn or severed tendons and ligaments, broken bones, and broken teeth as animals bite at the trap, sometimes biting away their caught limb in what trappers call a "wring off." Animals can be caught and held in a leghold trap for days, frightened, in pain, unable to eat or drink or care for their young. Although some states require trappers to check these traps every twenty-four hours on public land, other states do not. Very few states regulate how animals are trapped on private property.

Leghold traps are also used in the United States and Canada for other furbearing mammals such as coyotes, beavers, muskrats, minks, otters, raccoons, foxes, martens, fishers, and bears. Advocates say that leghold traps can be padded and allow trappers to release nontarget and domestic animals, who most often are not injured by the experience.

Other common traps are body-gripping traps, meant to kill an animal immediately by crushing the throat. These are often used underwater for mink, muskrat, and beaver. Unfortunately, when nontarget species—larger or smaller animals for which the trap is not designed—get caught in body-gripping traps, they can also be injured and held for days. One province in Canada allows trappers to check these traps every two weeks.

Snares are also used in trapping to catch animals around the neck or leg. For snared animals, the question, again, is how long they suffer before death.

The United States typically registers 250,000 trappers a year, a number that increases when fur prices are high. Canada has roughly 60,000 active trappers, of which 25,000 are indigenous people following cultural tradition. The majority of North American trappers are rural white people, many of whom also say they are following cultural tradition. According to America's National Trappers Association, trapping often provides supplemental income for workers in seasonal jobs like construction, farming, ranching, and logging. Government agencies sometimes use trapping to control animal populations that are damaging ecosystems, such as introduced nutria in America's southeastern wetlands. Conservationists sometimes use trapping to control species endangering other species.

The estimate of wild animals trapped and killed in North America each year ranges from six to ten million. Mostly this is about fur, which is then sold to the clothing industry. My friend Steve Macdonald is a wildlife biologist who grew up trapping in Minnesota in the 1950s, did subsistence trapping in Alaska, and later trapped small mammals—shrews, voles, mice, rats—for museum collections at the University of Alaska and the University of New Mexico. I have turned to him to try and understand this world.

Steve's father was a quiet man who partially supported his family by trapping weasels. Steve skinned his first rabbit when he was five years old.

Sitting on his porch with me, Steve describes walking behind his dad and recognizing—maybe still five or six years old—a long scat on the trail. Twisted like a rope. With undigested hair. "Hey, Dad, looks like a coyote." His father stopped, looked down, and said, "Good job." This was not a man who spoke much. Steve gives one of his deep laughs. "And I remember that today. My heart, you know, still swells at that memory."

Steve follows this with an emphasis: "My father never had a bad thing to say about animals. There were no good or bad animals." No villains, no pests, nothing to eradicate. Every animal had its place in the ecosystem.

When Steve trapped for a living, animals died quickly. That happens more quickly and easily in cold climates. He was careful not to overtrap animals whose populations were vulnerable. "You have to pay attention. For some species, it's not that hard to extirpate them." Many trappers today speak similarly about the importance of ethics and the role of science. But trapping, Steve says, is "now more chaotic." More people have recreational traplines, so that there are more traplines, managed differently in every state.

Steve wonders how trapping for income makes sense today given the cost of gas, vehicles, traps, GPS trackers, sometimes dogs, sometimes computers. I wondered that, too, when I read a 2022 market fur report that said beavers were expected to average $10–15, river otters $10–25, and muskrat $2–3. Wild mink were $3–5. Red and gray foxes were at a bottom of $5–10. The moneymakers were few and included bobcats. A top-quality bobcat could bring in $200–300. A top-quality Canada lynx could bring $50–100. Coyotes were suddenly, surprisingly, marketable because some high-end parkas were being trimmed with their fur.

Fur is actually a wonderful thing to wear for warmth and durability. "Nothing," Steve says, "beats fur." Moreover, as Melissa Kwasny points out in her book *Putting on the Dog: Animal Origins of What We Wear*, the use of natural materials like fur, leather, wool, and silk is less harmful to the environment than synthetic clothing that is made from plastic derived from oil. The production of synthetics is a significant contributor to global warming. The laundering and discarding of synthetic clothing also contribute to the microfiber pollution filling our oceans and waterways. Even natural plant materials have a high cost. The World Wildlife Fund estimates that the production of a single cotton T-shirt requires more than seven hundred gallons of water and a liberal use of pesticides. The overconsumption of clothes is a huge concern.

It's complicated. A world of eight billion people is not going to stop relying on animals for food and clothing. The pertinent debate then is how we get that food, how we get that clothing, how we treat domesticated animals, how we treat wildlife.

I'm sometimes startled by David Attenborough as he moves in his nature shows from rainforest to tundra to ocean floor. "I am traveling now in the Great Rift Valley. . . ." "But here in the Antarctica . . ." All, seemingly, in a day's work, never out of breath.

In imitation: here I am in a snowy field in Montana looking at a set of Canada lynx tracks. The snow is fluffy, and the tracks look nothing like those of a cat. Only if I focus on the bottom of these depressions can I see the feline's four tear-shaped toes, a leading toe, and no claws. Farther down this dirt road, the snow thins, and the tracks become clearer.

Although lynx are a rangier and lighter animal than bobcats, their feet are much bigger. The fur on these feet and the relatively long toes—which spread out in snow—create a flotation device or snowshoe. This helps the animal chase its main prey, the snowshoe hare, which also has large furred feet. Compared to bobcats, lynx have more prominent ear and cheek tufts, which serve as hearing aids for the soft footsteps of prey. The longer, grayer fur of a lynx is often without pattern and without spots on the belly. The longer hind legs give the animal a slightly arched rather than a straight back.

For trackers—for everyone—fresh snow is like someone covering your eyes and steering you into the living room where they have piled a mountain of wrapped gifts. All these gifts are for you. In snow, colors look new. The green is familiar but different against so much white. The expanse of white is almost an emotion. The white snow and blue sky are almost a mystical experience. There is transformation. Rocks have turned into pillows. Trees have become lampposts. There is the gift of your breath in the air. For trackers in the right kind of snow, there is the gift of perfect prints that show double lobes and trilobes and heel pads and teardrop toes and claws and everything else your tracking book mentions. There are multiple perfect tracks: lynx, foxes, mice, squirrels, turkeys, quail.

Janice Przybyl describes the experience as "cheating" because tracks in snow are often much more distinct and long-lasting than in other substrates. "A dusting of snow," she says fondly, "can change the most obnoxious surface for tracking, like concrete or rock, into a hologram of an animal's passage."

Snow also means weathered, windblown tracks that distort oddly as the snow melts and ice forms. In snow, animals move at unusual gaits. They might walk instead of trot. They use each other's tracks to save energy.

Bobcats will decrease their stride to fit into the prints of a porcupine or stretch their legs to fit the hoofprints of a deer. For obvious reasons, lynx like to use the trails made by snowshoe hares.

The number of lynx in North America is probably in the hundreds of thousands, especially during population peaks that occur with peaks of those snowshoe hares. Lynx are among the few predators that have such cycles. In Canada and Alaska, lynx are managed as a furbearing animal. They are hunted and trapped, and their population is described as secure or "apparently secure," although endangered in a few areas such as Nova Scotia and New Brunswick.

In the contiguous United States, the lynx is a threatened species under the Endangered Species Act. We don't know how many of these animals remain, maybe a total of hundreds, with small resident populations in Minnesota, Montana, Washington, Maine, and possibly Colorado. A 2018 report from the United States Fish and Wildlife Service, under President Trump's administration, suggested the species could be delisted. But reports by conservation groups assert that the lynx has not recovered from the overhunting and trapping of the twentieth century. Their populations are further threatened by habitat loss and fragmentation.

If there are lynx in Canada and Alaska, why should we care if they remain in their historic range in the contiguous United States? One answer is that a natural diversity of species is a measure of ecological health. Healthy habitat has value for humans, as well as its own intrinsic value. But for lynx, these good intentions are becoming less relevant. On a planet that is changing because of global warming, the ranges of animals and plants are changing, too. Inevitably, lynx will be pushed farther north not because we have delisted their status but because we have altered the Earth's climate.

IUCN MOUNTAIN LION, BOBCAT, AND LYNX ENDANGERED STATUS

Mountain lion, least concern, population trend decreasing
Bobcat, least concern, population trend stable
Lynx, least concern, population trend stable

Domestic cats can be divided into three categories. House cats live in houses, perhaps sixty million in the United States and eight million in Canada, often with the freedom to go in and out. Farm cats are a specialized group that live outside, typically in a barn, dependent on humans but tasked with keeping a building or area free of rodents. Feral cats also live outside, are not dependent on humans, and will not likely socialize to humans past kittenhood.

As many feral cats live in the United States as house cats, another estimated sixty million or more. Some wildlife biologists consider them to be one of our most destructive exotic species, killing billions of birds in a time when so many bird species are threatened. Feral cats often live short lives filled with hunger and disease. They are a huge, sad problem.

A cat is a cat is a cat, form following function, whether house cat, bobcat, or mountain lion. Occasionally the track of a large domestic cat might be mistaken for that of a small bobcat, just as the track of a large bobcat might be mistaken for that of a small mountain lion. There are subtle differences, however, in the width of the pad or the prominence of trilobing at the bottom of the pad. An experienced tracker would be flat on the ground, mumbling to herself.

Bobcat tracks are mostly smaller versions of mountain lion tracks. The asymmetrical track typically registers four toes, one middle toe slightly extended above the other middle toe. Claws are not usually present.

FIGURE 7.3 Bobcat, hind track

The fronts of a bobcat are about one and a half to two and a half inches, slightly larger and rounder than hinds. The palm pads are relatively large, holding all or most of the toes, with two lobes at the top of the pad and three lobes at the bottom. You can draw an upside-down C in the negative space of the front track and a C or an H in the hind.

House cats are smaller than bobcats, with front tracks about one to one and a half inches long.

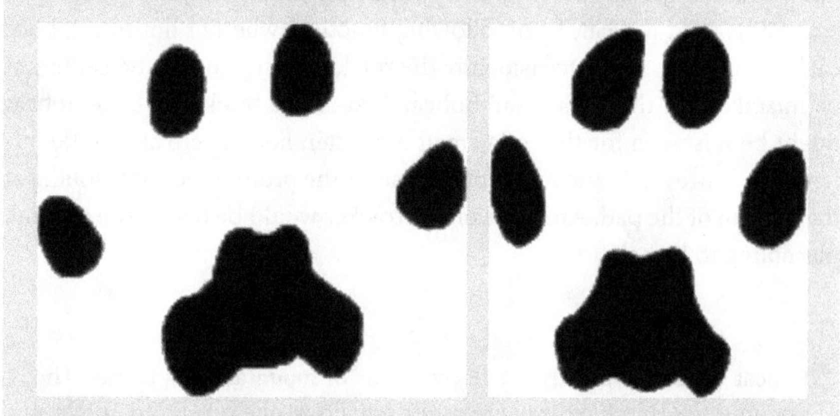

FIGURE 7.4 Domestic cat, left front and left hind

In a lynx, the abundant fur on the foot means lots of negative space between the toes and pad. Fur makes the palm pad look relatively small, not large—the opposite of a bobcat and other felines—and the long toes also look small.

FIGURE 7.5 Mountain lion, bobcat, coyote

WILDLIFE TRACK AND SIGN CONVERSATION #190

"Domestic dog."

"Domestic dog."

"Not a dog. Four toes. H in the negative space. Signs of fur. Trilobing at bottom of pad."

"H for gray fox."

"H for the hind of a small bobcat, which sometimes resembles a large gray fox."

"A bobcat pad would be relatively larger than a gray fox."

"*Relatively* is such a relative word."

"Also, the middle toes on the bobcat's hind track would look different from a gray fox's."

"How different?"

"I forget."

"The perennial problem."

8

A Black-and-White Aesthetic

My second class on identifying wildlife tracks was October 2018 at the Aravaipa Canyon Preserve in southeastern Arizona. Organized by Sky Island Alliance, the four-day workshop would certify trackers to participate in their wildlife monitoring program. About a dozen of us, mostly in our fifties and sixties, slept in the guesthouse or camped nearby. A few younger volunteers had come from a partnering program in Mexico. Those of us who knew Spanish enjoyed sharing dinners with them, talking politics, talking tracks.

In the evenings, we listened to lectures, going through the basics at speed.

Plantigrade: animals like skunks, raccoons, bears, some mustelids, and rodents walking on all the bones of their feet. A complete track shows toe pads, palm pad, and heel pad. Humans are a plantigrade species with five toe pads.

Digitigrade: canines and felines walking on their four toes or phalanges. Digitigrade species evolved from plantigrade species in order to move faster. Bones in their feet extended up, lengthening the leg and forming a higher and springier ankle joint. A complete track will show four toe pads and what trackers call the palm pad. Higher on the front legs, a reduced toe and heel, or carpal pad, can also appear in the track when the animal is moving fast or in deep substrate.

Unguligrade: animals like horses, deer, elk, and pronghorn that evolved to have even longer legs and run even faster on one or two hooved toes. A complete track will show the imprint of those hooves. Tracks can also show reduced toes, or dewclaws, higher up the legs.

We looked at skulls, big and small, as well as other bones from a kill site or natural death. We talked about how to document the tracks we found and why that was important. We heard about the jaguar and ocelot sign that some volunteers were seeing in Arizona and New Mexico—and how exciting that was. In the mornings and afternoons, we went out to Aravaipa Creek, where cottonwoods, sycamores, and willow shaded an intimate thread of water. We felt the reverence unique to those who live in the desert. *Sin agua, no hay vida.*

We saw all the usual suspects. Black bear tracks in the abandoned orchard near the guesthouse. Raccoon and coati along the creek. Mule deer and javelina. Domestic dog, coyote, and gray fox. A nice set of bobcat. A blurred and partial set of mountain lion. We talked about the problem of blurred and partial tracks. A mountain lion and a domestic dog track can be confused when a dog's foot has slipped in mud, distorting the pad, or when the claws of the mountain lion have extended to grip a slippery surface. The track of a domestic cat might resemble that of a spotted skunk that is not showing the heel pad, fifth toe, and claws. Both can look like a ringtail cat, which is not a cat but another relative of the raccoon and coati.

Of course, we saw skunk tracks, mostly striped, some hog-nosed. Unlike dogs and cats, plantigrade skunks have five toes on both feet. These toes typically register. The long digging claws of the front feet usually register. Both palm and heel pads often register. Skunk tracks sometimes look like small human handprints, which can be unnerving.

In their discussion of skunks, biologists use the phrase "honest signal." The white lines of a striped skunk point to its rear end, where two anal glands produce and hold about an ounce of musk each. This concentrated fluid can be sprayed with accuracy at targets ten feet away and with less accuracy fifteen feet away. The four other species of skunk in northern North America use the same pattern of white on black so as to be easily seen and avoided. This visual cue is enhanced when the skunk raises its tail.

Skunks don't want to use up their musk and then have to make more. They might instead eject a warning smell. They might hiss and stamp

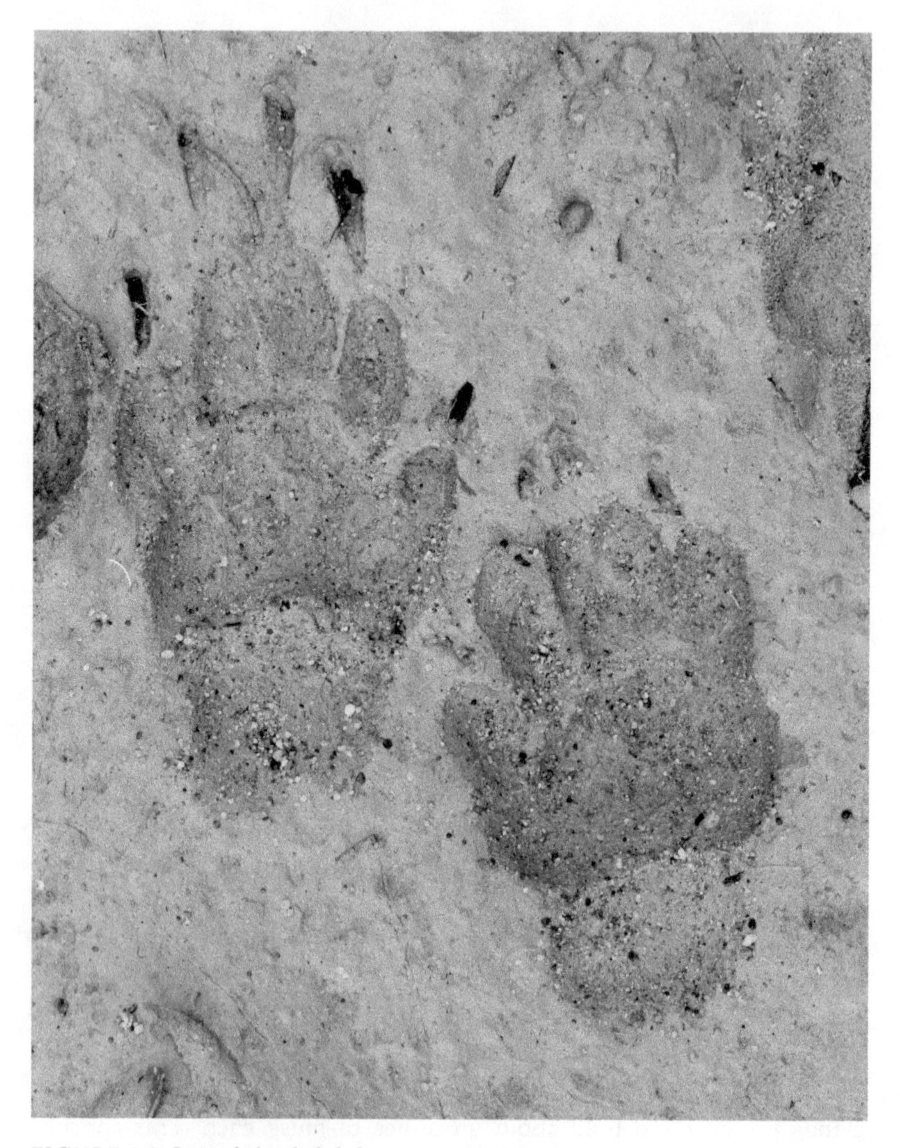

FIGURE 8.1 Striped skunk, left front and right hind

their feet. If these strategies don't work, if that domestic dog keeps barking and barking and coming closer, a striped skunk will turn so that both rump and eyes face the threat. From the rump, a stream or multiple streams of liquid can burn exposed skin and cause gagging or temporary blindness.

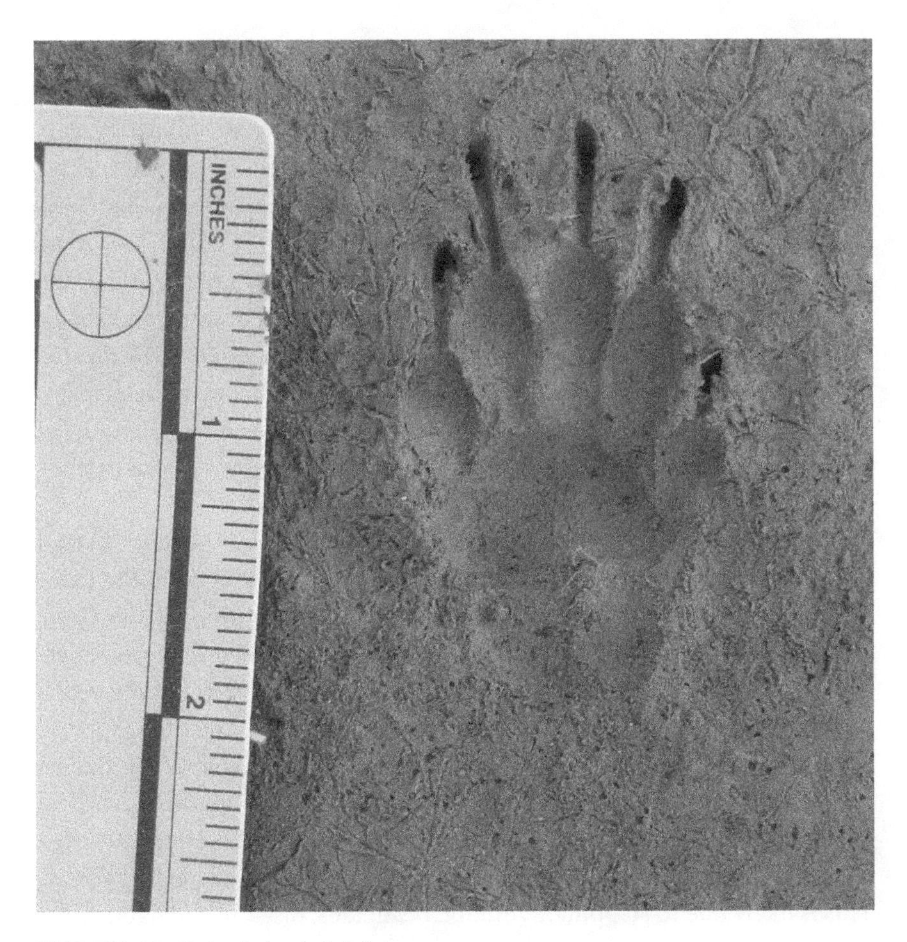

FIGURE 8.2 Striped skunk, left front

The result of such a strong defense is that skunks are relatively docile and, as one scientist says, "seemingly carefree." Adult striped skunks, weighing only a few pounds, have been videotaped eating from a carcass while a fox or mountain lion waits nearby. In one scene, three coyotes circled a skunk on top of a dead deer. The coyotes approached, retreated, split up, whined, fussed, nipped forward, nipped back. The researcher Stan Gehrt noted, "It was hilarious."[1]

The abundance of skunks in North America must be galling to other predators.

Striped skunks range from southern Canada to northern Mexico at an estimate of five to thirteen per square mile. They vary in size, weighing up to thirteen pounds. Typically they have a white stripe that begins at the ears, splits to form a double stripe down the back and joins again at the tail. But that pattern can vary, with some striped skunks mostly black, a very few mostly white, and some brown. Like all skunks, they are active at night. In cold climates, they enter a dormancy or deep sleep that slows their metabolism. These skunks den from November to March and must compete to find good dens. Housing is one reason they are moving to the cities and suburbs. They particularly like spaces under porches and sheds. They like to eat the grubs and insects in lawns and gardens. Dedicated omnivores, they also like small reptiles, small birds, small mammals, crabs, crayfish, eggs, fruits, roots, crops, kitchen scraps, and more. When I lived in Silver City, they commonly came in through the cat door because they like cat food.

Spotted skunks are smaller than striped skunks, with coats of partial stripes or spots over their back and sides. They tend to hide in the camouflage of vegetation. Their coloring warns predators who get too close, and they further intimidate by doing a handstand on their front feet, performing a split with their back legs, and walking forward in imitation of a flying carpet. Imagine that for a moment. Google it on YouTube. Spotted skunks also readily climb up and down trees. Western spotted skunks can be found throughout the western United States to southern Mexico. Eastern spotted skunks should be found throughout the eastern United States but are in serious decline and are listed as vulnerable by the IUCN. This is likely due to trapping, synthetic pesticides, disease, and habitat lost to agriculture.

The hooded skunk is confined mostly to the American Southwest and Mexico, with the variable coloring of a striped skunk but a longer tail and a fulsome ruff of fur around the neck.

The largest skunk, the American hog-nosed, also lives in the American Southwest and throughout Mexico. This animal has a powerful upper body for climbing in rocky areas, an elongated snout, and long claws for digging. The back of a hog-nosed skunk is often pure white, extending from the top of the head to a relatively short and thin tail. Unlike other skunks, the hog-nosed skunk has no white marking between the eyes.

My friend Harley Shaw once worked on a study of skunks, with images captured by game cameras. He says that neither he nor the other experts could comfortably identify striped, hog-nosed, or hooded skunks from photographs. Their color variants were just too varied.

More than half of wild skunks die in their first year. Those that survive might live another four or five years. Sometimes young skunks are killed by predators like bobcats, coyotes, and foxes while the mother is out foraging. Sometimes male skunks—who do not help raise their young—find and eat a litter. Captive females will also kill their litters if disturbed soon after birth. Great horned owls and eagles hunt skunks. Skunks don't understand cars at all. Mainly, though, skunks get diseases like rabies, distemper, and pneumonia, as well as parasites. Although rabid skunks rarely bite humans, skunks are a vector of rabies, which is carried in their saliva and often kills them in communal dens.

Tracking wildlife is about applying what you know to what you see. The details and geometry of feet. Animal behavior and ecology. Multifactor reasoning. It's about the virtue of patience—measuring that distance from claws to the bottom of the palm pad one more time—and humility, remembering the last time you spoke too soon. Above all, it's about curiosity. Are these the partial tracks of a large spotted skunk or a small striped skunk? Patient and humble observation might tell you which, but are you that curious? Do you want to spend that much time? Of course, you will become a better tracker the more you practice, and getting better is a strong motivation. But you also have to care, really care, about what animal was here a day or so ago.

The truth is that sometimes I care, and sometimes I don't. I am sometimes pleased to just say "skunk," which is a genuine pleasure since I like skunks so much. My next thought is about lunch or something I need to do after lunch. This limited curiosity will always limit me as a wildlife tracker.

I see skunk tracks all the time, every day, around my house, in the fine substrate of the road that leads up to the irrigation ditch, in any bit of dirt or mud, and in the garden enclosed against javelina but not skunk. Skunks regularly explore our wraparound porch, so that my husband and I watch the animal out one window and then hurry to see it from another. Skunks are often under the bird feeders. We smell them in the morning—the result of some interaction with another animal. We've startled them in the early evening. They jump. We jump. Once, jumping back, a hooded skunk fell off the rise of the porch onto his back. Chagrined, he waved his paws in the air like a baby or a pill bug before more calmly turning over and galumphing away. These skunks have never sprayed or shown us any threat behavior. We don't have any pets for them to worry about. If we live companionably with any wild mammal, we live companionably with skunks.

FIGURE 8.3 Striped skunk
(photo by Elroy Limmer)

Sometimes we see skunk families, mothers and juveniles. I have seen a mother determinedly waddle and forage while an addled youngster, also determined, dragged under her belly still trying to nurse. The males finally disperse at three months or so. The females might stay for almost a year. Mostly we see the solitary skunk. It's like a fashion show. Something from Chanel. A bit of art deco. That black-and-white aesthetic. An honest signal flashing in the night.

IUCN SKUNK ENDANGERED STATUS

Striped skunk, least concern, population trend stable
Western spotted skunk, least concern, population trend decreasing
Eastern spotted skunk, vulnerable, population trend decreasing
Hog-nosed skunk, least concern, population trend decreasing
Hooded skunk, least concern, population trend increasing

Bryon Lichtenhan was one of my instructors at Arivaipa Canyon at the Sky Island Alliance workshop. Tall, bearded, in his late thirties, he had come to tracking as a homesteader and permaculturalist and through his interest in ancestral skills—the art of making clothes from deer and baskets from willow. Hunter, flint knapper, basket maker. These are still resonant occupations.

Ancestral skills have a loyal following in the United States and Canada, with more than forty gatherings a year called by names like Winter Count, Echoes in Time, Rabbitstick Rendezvous, and Stone Age Primitive Arts Festival. The theme may be obvious, but one website warns, "We are a diverse group from many experiences, opinions, and family backgrounds . . . you may find yourself around people who don't share your political, spiritual, or life views." You may indeed. Many people flint arrowheads and grind mesquite beans as a form of emotional health, a way to live in the modern world. But others track, skin, and wear animals as survivalists, preparing for the end of the modern world.

I like the jubilance of the Florida Earthskills Gathering. "We play our own music, create our own games, weave our own dance, make our own reality, and live our lives with feet firmly planted on the ground. We grow food, we regard water and air as precious resources, and we honor and respect all beings sharing this existence on Earth. We seek to become a healing part of all that is good and beautiful about life on Earth, and to share this ecological awareness with others for generations to come."

More staidly, the Buckeye Gathering in California intones, "We cannot roll back time to a pristine past, but we may learn fundamental lessons from the people who have come before."

Wildlife tracking classes and teachers often include this connection to the "pristine past" of the hunter and gatherer. The well-known tracker Tom Brown Jr. says he learned his skills from an Apache elder and shaman. A tracking website or organization might use a Native American term such as *vision quest*. But most trackers also have a twenty-first century awareness of cultural appropriation and refer to the heritage of tracking only with the goal of acknowledging their debt.

When Bryon learned that I live near the Gila National Forest in southern New Mexico, he asked if I knew Doug, famous on the ancestral skills circuit. Of course, I know Doug, who has traveled up and down the Gila River for more than a decade. I see Doug on trails in the national forest and occasionally at parties. Bryon asked next about a man called Wind. Yes, Wind is an outdoor neighbor of mine who lives in a brush shelter, makes his utensils from driftwood, and grows corn and beans in a nearby garden.

Some years ago, he came to a potluck dinner where he and my father-in-law, then in his late nineties, sat and talked about World War II. Mostly my father-in-law talked. Both Doug and Wind often go without shoes, and their barefoot plantigrade tracks are familiar to me.

For this tracking class, Bryon had an assistant, Rosemary Schiano. There's so much to teach in only a few days, most of that outside. You need someone to watch over the stragglers. Rosemary was about fifty, a muscular woman with a blond ponytail who had grown up on Long Island, New York. Her relationship with animals had started when she was two years old. Since that childhood, the numbers of more than five thousand animal species on Earth have dropped almost 70 percent, a statistic I used at the beginning of this book, and one that Rosemary may first have told me. Rosemary thinks a lot about defaunation. Witnessing the defaunation of America has become part of her life's work as a biologist. When I met her, she was moving back and forth in her jeep from a seasonal job with the United States Forest Service in Colorado to private research in Arizona, where she was documenting how the human presence—immigrants, Border Patrol, ranchers, tourists—affected wildlife in the Sonoran Desert. She estimated she was in the field then three hundred days a year.

Today, Rosemary gives lectures and training on how to coexist with predators, how to discourage mountain lions from eating your pets and livestock, and how to properly use bear spray. She is pragmatic. Human beings have taken over everything, everywhere, and there's no turning back. Still, she's always on the side of the animals. She's always explaining that we have to change how we treat wildlife. Certainly that means no trophy hunting, leghold traps, and lethal predator control. But also—and now she's going to offend people—any form of feeding wild animals. This includes bird feeders. Deliberately feeding wild animals is painfully misguided, an interference in their lives. Inadvertently feeding them, not covering your compost and trash, is negligent. Running or bicycling in landscapes with mountain lions is irresponsible, since this might trigger a mountain lion's response to prey, which would be mutually terrible. You can hike, she says. You can track. But don't run.

On paper, Rosemary sounds like a purist. I think we need more people like her. I think this even though I have bird feeders and run in landscapes with mountain lions.

In person, at the tracking workshop, Rosemary was easygoing and kind, which some of us adult learners needed. We had been trained to be good students, even the "best student," but we were getting too much information too fast. Moreover, there's an annoying pedagogic style in tracking workshops of withholding the answer until you've stretched your capacity to its limit—and then you still don't get an answer.

"So, it's not a skunk?"

"Or a raccoon?"

"Not a coati?"

"What about a gray squirrel?"

Argh. More than once, Rosemary stood at the fringes of the group, whispering a hint. "Look at the toes more closely." And, shhh, "Do you see any fur in the track?"

The best thing about a tracking workshop is being with trackers like Rosemary and Bryon. You can't really learn tracking from a book, as Rosemary would be the first to say. At the same time, experiential learning has to be supported by a lot of knowledge. At one point, she advised, "Just read and read about animals. Just learn and learn. Get them in your head. Look at the pictures. Remember size. The diameter of scat. Get all that in your head. It will come back to you on the trail."

On the second day of our tracking workshop, Bryon demonstrated gaits. A gait is how an animal moves. Walking. Trotting. Galloping. These movements result in consistent track patterns on the ground. For some of us, visualizing gaits and their track patterns can be the hardest part of identifying track and sign.

A walk seems simple, each foot moving independently, one side of the animal and then the other side, never losing touch with the ground. But then there are understep walks, direct or double-register walks, and overstep walks. Bryon showed us the difference by sweeping an area of soft dirt, arching his body into a V, and walking across as a four-legged animal. His hands were the front feet, his feet the hind.

In the understep walk, he moved slowly, as a stalking bobcat might, right hind foot landing behind right front foot, left hind foot landing behind left front foot. After we rushed over to see, he swept the area clean again. Next he walked a little faster, as a beaver might, in the double register, hind prints directly covering front prints. He swept the area clean again. In the overstep walk, he moved faster still, like a bear.

The increased speed meant that the hind foot swung more vigorously forward and landed ahead of the front track.

To determine some gaits—like the common overstep walk—you need to know that front track from the hind. The clues are often size and shape. Visualize the animal. The weight of a German shepherd's head and shoulders means its front feet are slightly bigger than its hind. A kangaroo rat that can move bipedally, hopping on hind feet, will have hinds twice as big as fronts. A ground squirrel that digs a lot will have heavier, longer claws on the front feet.

Sometimes you want to know whether a track is the animal's right or left. The position and length of toes can tell you that, just as they do on a human's hands and feet. Also, in a walk, when you have a group of tracks and can see a trail width—the distance between the right and left track—the right foot is always on the right side of the trail and the left foot on the left side. Typical trail widths have been measured for each species. This is one more thing for trackers to memorize.

The more quickly the animal moves, however, the smaller the trail width. In a trot, the trail can be just a single line, without a left or right side. Now Bryon did a trot, the natural rhythm of canines, felines, fishers, deer, elk, moose, voles, and short-tailed shrews, with the front and hind legs of opposite sides moving simultaneously. Trackers suggest you think of these two legs, diagonally opposite, as joined by a cable. Bryon looked quite athletic attempting a movement for which he did not have four legs. A trot might result in the double-register track, hind feet landing where the front feet did. A faster side trot is preferred by canines, who angle their body right or left as they move, the hind track appearing slightly ahead and to the same side of the front track. In a straddle trot, preferred by gray foxes and some ungulates and several shrew species, the animal kicks out the hind legs to alternate sides, with that pattern showing as a zigzag. Amazingly, Bryon was able to do these, too, his feet a blur, the concept humanly unnatural.

Bryon grinned and tried to mimic a lope and then a gallop, although he couldn't possibly achieve the two occasions in a gallop when the animal is airborne. Lopes and gallops have particular track patterns and come in various forms. Gallops are faster than lopes, with that greater speed showing again as hind tracks ahead of front tracks. Some people are better at three-dimensional kinetics than others. But watching real animals gallop, lope, and trot helps. Imitating an animal yourself is both instructive and admirable.

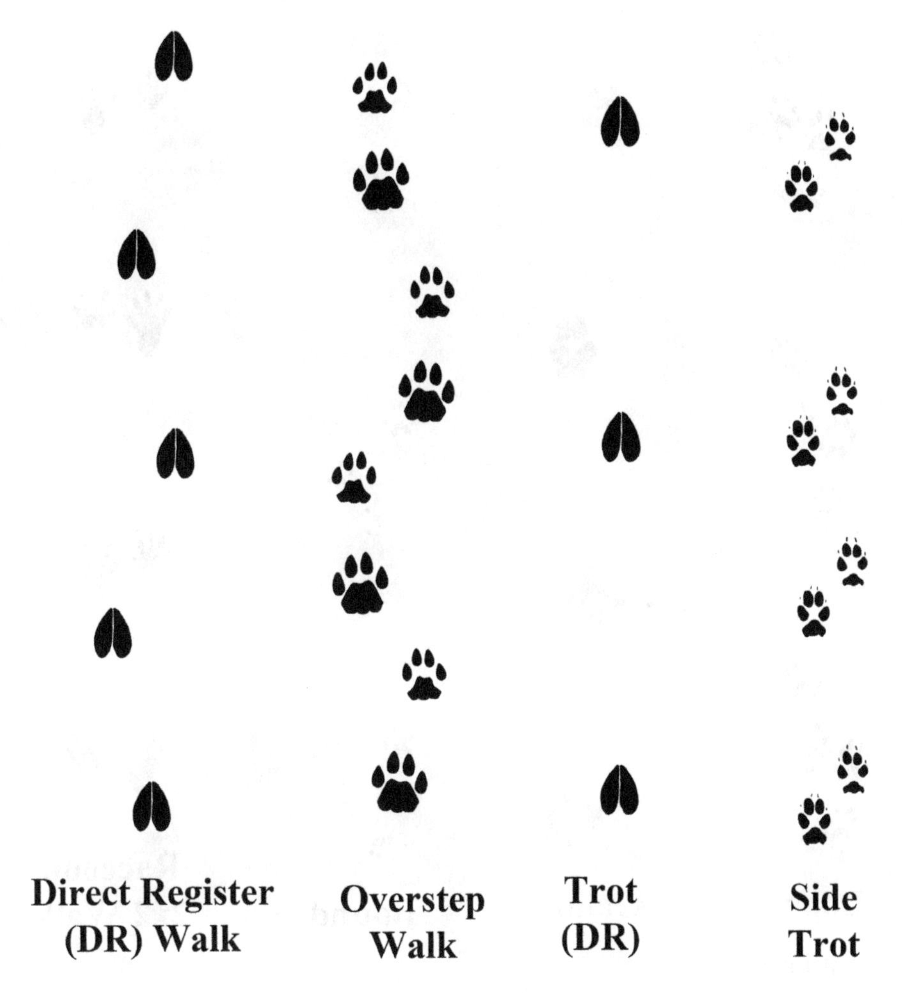

FIGURE 8.4

If Bryon couldn't do a gallop, he also couldn't pronk like a mule deer, pushing off from four feet and landing again on four feet. Still, he concluded his demonstration triumphantly. First he bounded like a rabbit or squirrel, front feet first and hind feet moving next to land on either side of and above the front feet. This was somewhat successful. Then he hopped like a frog, landing with two front feet above two hind feet. A man hopping is a fine sight and, by this point, we were cheering. Like a prince, Bryon stood up and took a bow.

Knowing the common gait of a species can help you determine which animal made this track. The common gait of a spotted skunk—the most

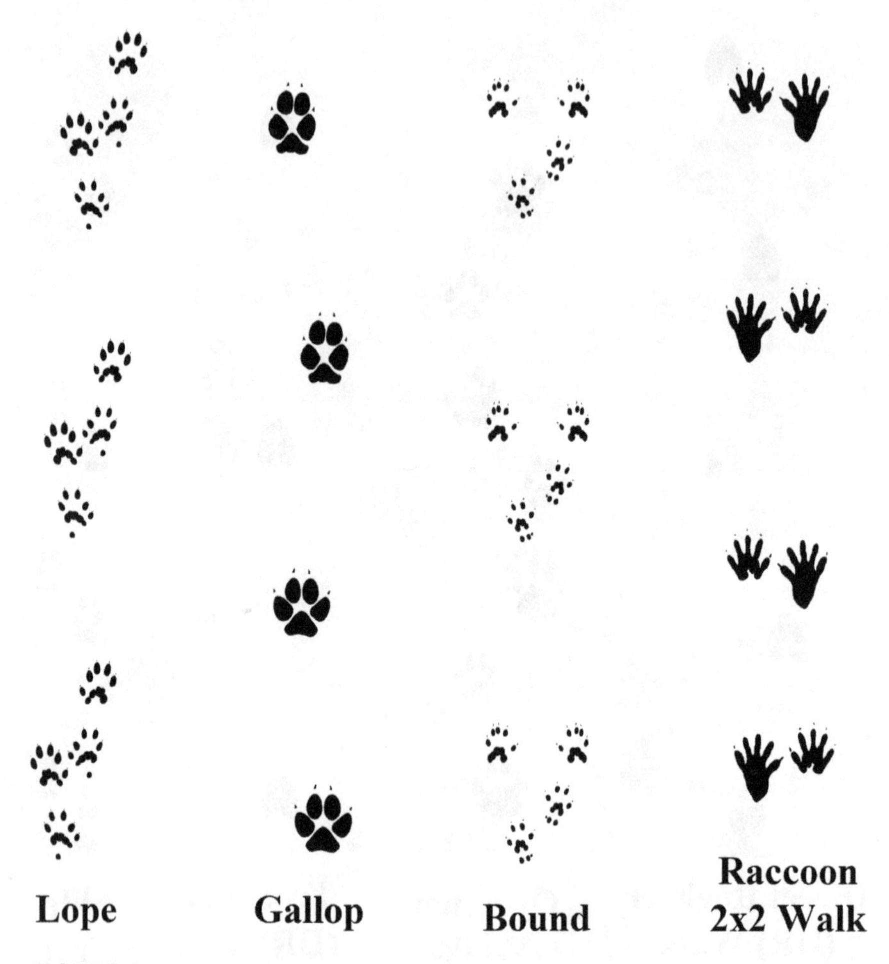

FIGURE 8.5

carnivorous of all skunks—is a bound. Seeing the tracks of a bound in the dirt might make you decide that this is a large spotted skunk and not a small striped skunk. Hooded, hog-nosed, and striped skunks more commonly walk. With short legs and a flat-footed emphasis, they tend to waddle. All four species lope. Around my house, skunks frequently break into a loping, waddling lurch, which is distinctly comic. Skunks can also gallop, as fast as ten miles an hour.

"Oh, skunks," Rosemary nodded. We were ambling down Arivaipa Creek, October 2018. The cottonwoods wove a yellow bower. The air smelled of water and leaves. "Who doesn't like skunks?"

Striped skunks usually register five toes, but not always. The smallest toe, or Toe 1, is in the place of our thumb, facing the inside of the track. This helps determine if a single track is a right or left foot. The toes do not splay. The claws of fronts typically register and are longer than the hinds, which is generally true of digging animals. Tracks range from slightly more than one and a half to slightly more than two inches. In front tracks, the palm pads are fused to make a single trapezoidal shape. The round heel pad may not register. In hind tracks, the small Toe 1 and shorter claws may not show. The toes do not splay. The palm pads are fused to make a single large pad. The heel pad of a hind track often registers and is separated from the palm pad by a distinct seam.

Hooded skunk tracks look a lot like striped skunk tracks.

The tracks of the hog-nosed skunk, an avid digger, are larger, with longer front claws. Front tracks can be two and three-quarters inches and hind tracks two and a half inches. As with all the measurements in this book, these are taken from claws to the bottom of the heel pad.

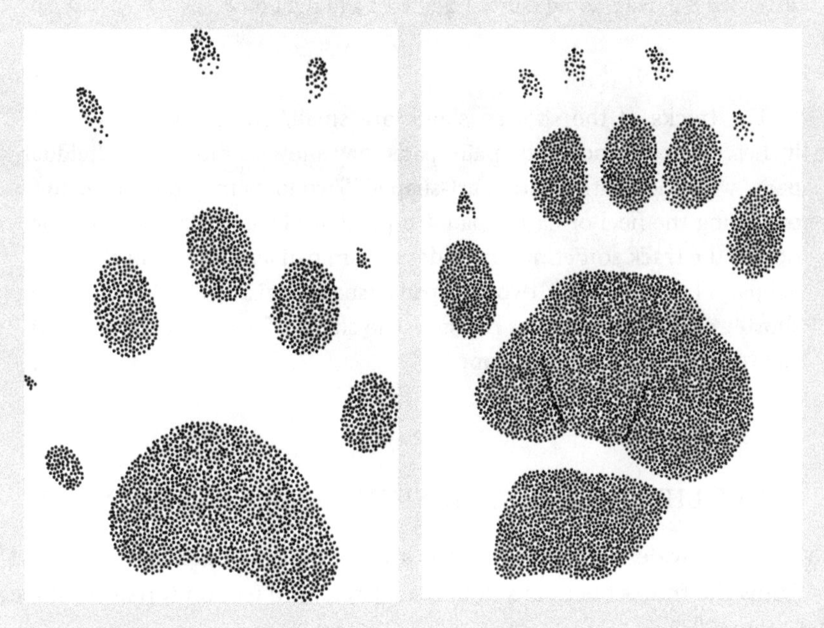

FIGURE 8.6 Striped skunk, right front and right hind

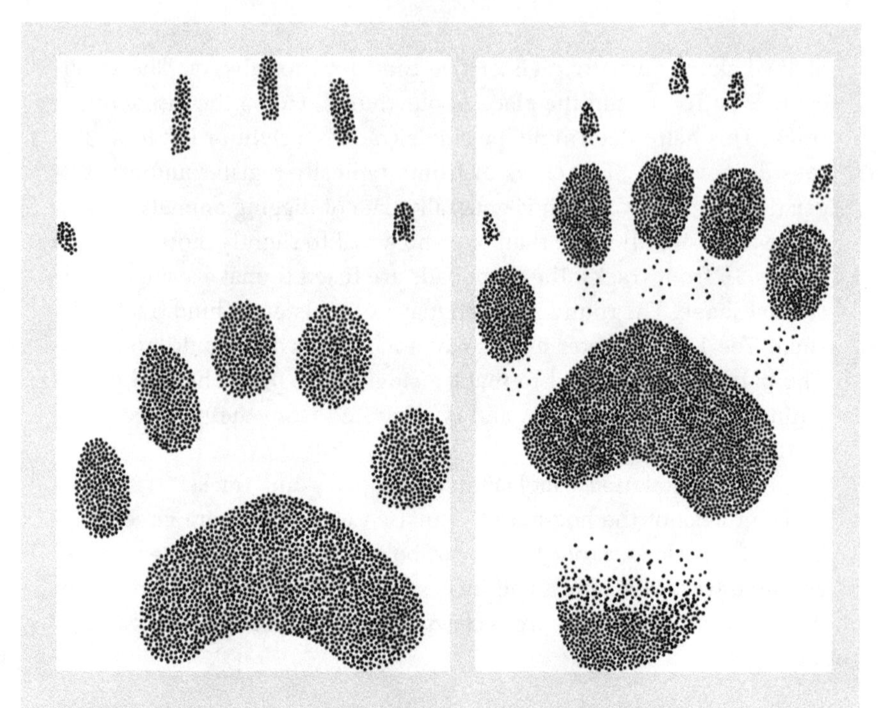

FIGURE 8.7 Hog-nosed skunk, right front and right hind

The tracks of the spotted skunk are small, one to one and a half inches. In front tracks, the palm pads may show as multiple individual pads, with the central pad heart-shaped. Two pads may register below, one being the heel or carpal pad. (In plantigrade animals, the heel portion of the track sometimes includes a palm pad as well as a heel or carpal pad.) In hind tracks, two heel pads usually register. Hind tracks have shorter claws that may not register. The shape of a spotted skunk track can resemble an ice cream cone.

WILDLIFE TRACK AND SIGN CONVERSATION #61

"It's a pretty wide trail, which means a walk, front feet ahead of hind feet."

"Now the tracks become a double register, hind feet on top of front feet. A faster walk."

"The trail width narrows here to a straight line. This is a trot."

"Now a side trot."

"Coyote walking down the road, walking faster, breaking into a trot, then a side trot."

"Sees something to eat."

"Smells something to eat."

"Feeling frisky."

"Feeling alive."

9

The Entangled World of Humans and Raccoons

An animal is creeping around my house at night, walking quietly, in starlight and moonlight and under the black depths of clouded skies, unseen—at least, not by me. This has been happening for years now, that visitor crossing the irrigation ditch and coming down the driveway, invisible but for the tracks I find the next morning. It's not just a specific animal, either. It's an entire species, seemingly shy, unobtrusive, careful to stay on the edges of human life. My friends here in the Gila Valley say the same thing. Like me, they have seen a lot of wildlife over the years. Coyotes, gray foxes, black bears, mountain lions, bobcats, skunks, coatis, ringtails, javelina, deer, elk. Ground squirrels and tree squirrels all day long. Songbirds and ground birds and waterbirds and raptors all day long. Lizards and snakes all day long—in the summer. But we hardly ever see raccoons. Only their tracks on the banks and in the shallows of the Gila River, as well as regularly, almost every night, by my house.

It's weird. Raccoons are better known for being gregarious and bold. In urban areas, they can be found in attics, sheds, garages, ventilation ducts, any good place to defecate or get together with friends and make noise. Or raise a family. More bright-eyed raccoons. In city parks and preserves, they are often conspicuous, peeping from tree holes, loping across paths. Commonly they are the despair of urban biologists because they reduce the population of native frogs and other amphibians, whose eggs appear early in the spring before raccoons can switch to better sources of food, such as the leftovers of human picnics.

In some cities in the United States and Canada, the raccoon population is more than two hundred animals per square mile. Toronto, in particular, has been called the raccoon capital of the world, with raccoons seemingly in every yard and at every kitchen door. One scientist believes that urban landscapes are selecting for an increasingly intelligent animal, an uber-raccoon, extremely good at opening doors and untying knots. Another scientist studied changes in weight and blood sugar levels based on where raccoons fed, from the food waste at the Toronto Zoo or in a neighborhood with more moderate garbage. The researcher chose raccoons because the animals seem to be everywhere. "Who among us," he explained, "hasn't seen a raccoon emerge from a garbage bin with a pizza crust in its mouth?"[1] Or watched a raccoon "for the better part of an evening wear a Dunkin' Donut bag on its head"?[2]

The raccoon's embrace of garbage is in contrast to the behavior of coyotes, who continue to feed on their more natural prey, and skunks,

FIGURE 9.1 Raccoon, left front and right hind

who are less physically able to get into bins and dumpsters. If you are studying such disheveled garbage, perhaps strewn across your yard, you might be struck by a physical memory of actually shutting and locking the mechanism on that trash can lid. You might look around now at any nearby dirt or substrate that would hold a track. Wonderfully often, the shape, size, and robust quality of five long toes practically scream "raccoon." The fingerlike toes are usually connected to the palm pad. The claw marks often show. Front tracks can resemble small human hands. Hind tracks can look like small human footprints.

Raccoons also have a recognizable walk, which is named after them, although grizzly bears sometimes use a version of this gait. In the raccoon walk, the animal moves the front and hind foot forward on one side of its body, then the other front and hind foot forward. This is a walk that even I can easily duplicate. The tracks on the ground show a front foot next to the opposite side's hind foot, a hind foot next to the opposite side's front foot, and so on. Again, there's a happy ping of recognition. You know this walk. You know this track.

The biologist Stan Gehrt, who has been studying urban wildlife for thirty years, marvels at what he calls the raccoon's neophilia. "For many other wild animals, when there's a strange object out there, they have a healthy fear of that. But raccoons are actually attracted to new, novel objects, shiny objects, things that are not normal in the landscape."[3] He might explain the stealthy behavior of raccoons in my valley as cultural, something learned by raccoons living near people who readily kill them. For country raccoons, shyness can be a good survival trait.

Raccoons also highlight the individuality of animals. While some raccoons in Stan's studies become hard to trap, others go willingly into the cage, curling up to sleep, time after time, as if enjoying this shared research experience. Raccoons do have patterns of behavior, but as Stan

FIGURE 9. 2

told one interviewer, "Typically you'll find 10 to 15 percent that will do the opposite."[4]

After the carnivorans diverged into dogs and cats, the animals that would become raccoons further diverged from the canid branch, as did skunks, bears, weasels, badgers, and otters. Once again, animals moved back and forth across the Bering land bridge and Panama isthmus. The ancestors of the northern or common raccoon first evolved in Central America and migrated to North America about four million years ago.

More than a dozen subspecies of raccoons range through southern Canada, all of the United States except the hot deserts of some states, and all of Mexico. Most government agencies don't bother to guess how many raccoons they have. Like skunks, foxes, and coyotes, there are just a lot. Minnesota alone may have up to a million raccoons, of which trappers and hunters kill about a quarter million each year. The IUCN lists them as of least concern, with population increasing.

In the wild, raccoons usually survive one to three years, killed by trappers and hunters, other predators, cars, or disease. They are more social than we once thought, males forming small groups and females the same, with females raising the kits alone until the juveniles leave at about six months or later. They flourish as a species largely because they are so flexible—about food, denning sites, breeding seasons, and mating. Global warming will only help these omnivores expand their range.

In some places, they are an invasive species. First introduced into Germany in the 1930s, the animals are overrunning European cities, where they behave just as badly as they do in American cities. The Germans catch and kill them in chocolate-baited traps. Sightings in Spain and Scotland have caused alarm, with calls for "urgent action." In Japan, a popular book in the 1960s and a long-running cartoon show resulted in thousands of young raccoons being imported into the country as pets. When the juveniles matured and became troublesome, thousands were released into the countryside, where they destroy crops and the fragile wood of Buddhist temples. Just as Americans despise nutrias and kudzu, other people despise raccoons.

But raccoons are in books and cartoons for a reason. Their masked face is delightfully expressive, alert to life and the wide world. Their manipulation of objects is especially familiar as they use their front paws to probe, pry, extract, and appraise. They are often seen dabbling intently in a stream or pond, their sensitive fingers searching and feeling underwater. Eventually the animal lifts a crayfish, a snail, or a plant to its mouth. (Captive raccoons dabble instinctively, which led to the false idea that raccoons wash their food before eating.) We understand this kind of exploration. We also understand going straight for the doughnuts.

An experiment called the Aesop's fable test, from the story of a thirsty crow, is designed to measure an animal's grasp of cause and effect. Eight raccoons were presented with a cylinder that contained a low level of water and a floating marshmallow. First the scientists demonstrated the trick of dropping stones to bring up both the water level and the treat. Two of eight raccoons watched carefully, dropped their stones, and got their marshmallow, as do crows and great apes. A third raccoon climbed onto the cylinder, rocked back and forth, and tipped everything over. He also got the marshmallow.[5]

A paper in the journal *Environmental Humanities* by Veronica Pacini-Ketchabaw and Fikile Nxumalo describes raccoons and their "wily Trickster ways" as a necessary part of "an entangled multispecies world." In this case, the animals were entangling with children and teachers at a university childcare center on Canada's west coast. For hundreds of years, raccoons had lived in the forests that predated the university and town, their intelligence recognized by the people who also lived in these forests. Trickster Raccoon had become part of many indigenous stories that taught about respect and cooperation among animals and between animals and humans. Sometimes the story was cautionary: see what happens when you steal food. Sometimes the story was celebratory: see what happens when you steal food.

The authors saw raccoons as tricksters still, crossing boundaries and causing trouble. At the childcare center, raccoons crossed the boundaries of a "perceived nature/culture divide," or the mistaken belief that humans are separate from nature. Literally, raccoons tried to enter the classrooms through open doors and windows. They made dens in the storage sheds. They played on the playground equipment. Psychically, they crossed an "ontological divide" when their curious, expressive natures made them seem humanlike, entertaining the children but also creating anxiety and surprise. Finally, they crossed the human/nonhuman divide with their microbes, with some parents and teachers worrying about rabies, others about roundworm.

Like many animals, raccoons transmit disease. Every year, between thirty and sixty thousand people in the United States get prophylactic treatment against rabies, usually associated with a bite from a bat, a raccoon, a skunk, or a fox. (Dogs and cats carry rabies, too, but most domesticated pets are vaccinated.) Every year, one to three cases of actual rabies are reported in humans, with an average of two deaths. Bacterial diseases like leptospirosis, listeriosis, and tularemia can also pass from raccoons to humans, although reports of these are rare. The raccoon roundworm is picked up from contact with raccoon feces. Again, documented cases are fewer than twenty-five in the past fifty years. But the parasite can be fatal, and less severe cases are probably undiagnosed and underreported. All tracking books warn about raccoon roundworm and suggest wearing gloves and washing hands when handling scat. Trackers also mention hantavirus in rodent scat and the bacteria leptospirosis in some mammal urine and feces.

Despite their concerns, the humans at this childcare center valued their relationship with raccoons and tried different ways of living together. They installed raccoon-proof windows. They educated themselves about respectful behavior around wild animals. They practiced drills that warned children when to give the raccoons "space." They checked constantly for raccoon feces, with the raccoons helpfully using a communal latrine away from the playground.

One morning, the resident family of raccoons, a mother and her juveniles, were found dead. Likely they had been poisoned by a campuswide program to deal with the growing raccoon problem. Another family of raccoons took their place, and these raccoons were even more aggressive and troublesome. Again, they were tolerated, sometimes happily, sometimes not. The authors concluded that the entangled world of raccoons and humans is "replete with charm, curiosity, and uneasy, asymmetrical, and tenuous relations."[6] Living in such an entangled world can feel both awkward and hopeful.

Raccoons typically show five long toes that connect to the palm pad. In front tracks, T1 and T5 are at about the same level, also called having the same "setback," and the heel pad sometimes registers. Claws sometimes register. In hind tracks, the smallest T1 is set farther below the other toes, and the heel pad often registers. Claws sometimes register. Front

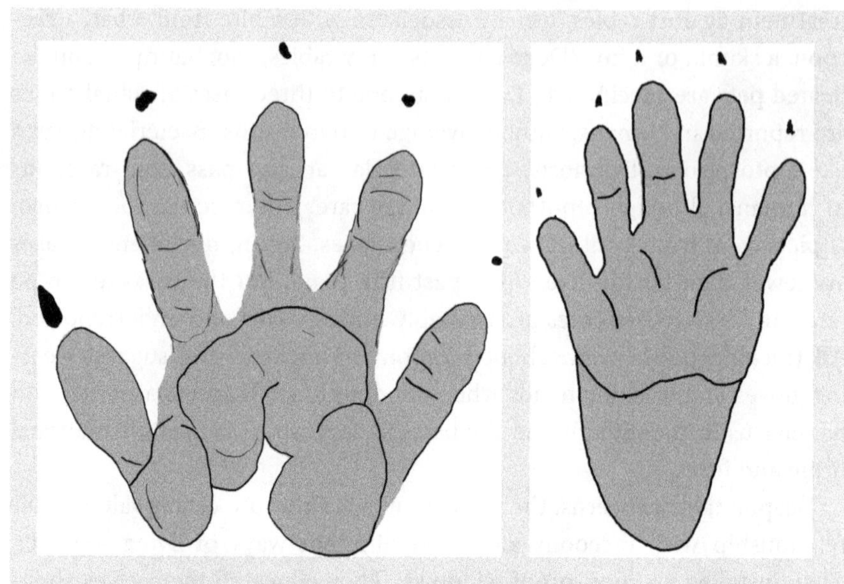

FIGURE 9.3 Raccoon, right front and right hind

tracks of raccoons measure one and three-quarters to slightly more than three inches long. Hinds are somewhat larger.

Like all animals, raccoons leave other sign. Commonly, raccoon scat will be with other scat in a raccoon latrine under a big tree or rock overhang. Animals who ingest and die from raccoon roundworm—mice, birds, rabbits—avoid these areas. Rats, who suffer only mildly from the parasite, will rummage through raccoon scat for undigested seeds. Lizards and bobcats, unaffected by roundworm, treat the latrines indifferently.[7]

In the same family as raccoons, white-nosed coati have been extending their range north from Mexico and Central and South America. In the United States, they are found in the woodlands and riparian areas of southern Arizona, southern New Mexico, and southern Texas. Arizona believes it has enough coatis for each hunter to kill one a year. New Mexico does not allow hunting. In Texas, the coati is on the state list of threatened species.

Figure 9.4 White-nosed coati
(photo by Elroy Limmer)

More than any wild animal, the coati has crossed my psychic boundary. When I see a coati, I feel suddenly in the middle of a children's story in which animals talk and advise the main character. I can't explain why. Yes, coatis are charismatic with their striking coats and facial markings, their social complexity, their range of sounds—the absent-minded grunt, the trilling chirp, the loud ha-ha-ha! But many animals are just as interesting. Perhaps it is that tribal matriarchy. All those females telling everyone what to do. Perhaps it is because coatis are diurnal, chittering and huffing to each other in the morning and late afternoon when I am also out and about—adding up to a dozen encounters in as many years. Perhaps it's the long nose. The long, banded tail. As the poets say, the heart wants what the heart wants.

The third relative of a raccoon that lives in North America is the ringtail, the size of a squirrel, about three pounds, with a catlike body, foxy face, protuberant eyes, big ears, and another beautifully banded black-and-white tail. Not often seen, ringtails like to climb trees and live in high places. They are spread broadly throughout the American West, up into southern Oregon and across to Kansas, and south through much of Mexico. They prefer rocky areas and have adapted to aridity, hunting at night and physically able to conserve water. The IUCN notes that ringtails are "found always in low densities and considered not abundant."

In Gold Rush California, miners kept ringtails as a way to control mice in the cabin, and the animals came to be known as miners' cats or ringtail cats. They are the state mammal of Arizona, although few native Arizonians (such as myself) have ever seen one. Their average life span in the wild is seven years, surprisingly long for a small mammal. The males have been known to bring food to pregnant females and play with their young. When threatened, ringtails scream and exude musk. They are the most carnivorous of their relations.

My friend Sonnie yearned to see a ringtail track and so she did, all the time, and then had to correct herself. Like raccoons and coatis, ringtails have five toes, but their smallest toe doesn't always register. Their semi-retractable claws don't always register. Their distinct palm pad is often smudged. For these reasons, the track of a domestic cat or a spotted skunk might look like that of a ringtail.

"I really think this is ringtail," Sonnie would say. With a little difficulty, she got on her knees. Studying, nodding, pretending to think, she spoke with a certainty we usually avoided. "This looks like a ringtail. This is ringtail." The animal had crossed her psychic boundary.

IUCN RACCOON, COATI, AND RINGTAIL ENDANGERED STATUSES

Northern raccoon, least concern, population trend increasing
White-nosed coati, least concern, population trend decreasing
Ringtail, least concern, population trend unknown

WILDLIFE TRACK AND SIGN CONVERSATION #78

"Three long thick toes pointed forward and one backward."

"Turkey tracks."

"A lot of tracks. A group of females?"

"Taking a dust bath."

"Over here, raccoon."

"Perfect set of tracks."

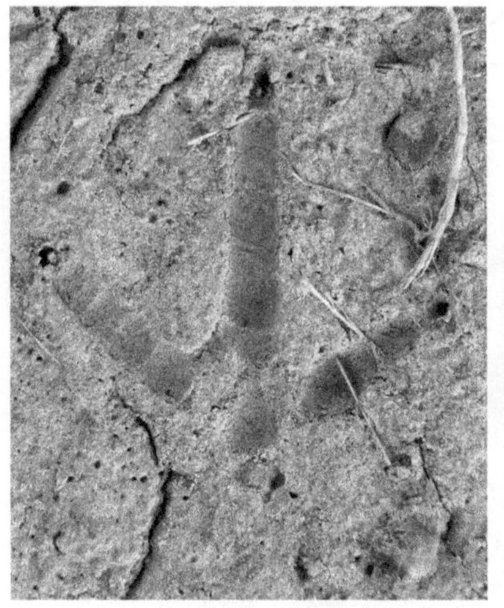

FIGURE 9.5 Turkey and raccoon tracks

"Beautiful set of tracks."

"Sometimes you know enough to know you don't know. Once in a while, it's nice to know enough to know you do know."

"Turkeys and raccoons."

10

Three Bears

I am always missing bears. "Did you see the bear that walked past the pond?" "No, I didn't see it!" "You just missed it!" Once I found a pile of bear scat warm to the touch, the diameter of a dessert plate piled high with digested juniper berries. In a few minutes, another hiker came toward me. "Did you see that bear?" "No." "Just missed it!"

And people are so cheerful, like this was a traffic light, not a bear.

After the workshop at Arivaipa Canyon, after Bryon and Rosemary took us to the bear tracks in the orchard, I went home and saw bear tracks up and down the trails I normally walk and run. Somehow I had missed them before. About this time, I also started accompanying Sonnie and other volunteers for Sky Island Alliance on their transect in the nearby Burro Mountains. We saw a lot of bear tracks. Most were iconic, five toes with the small toe (like your thumb) on the inside. Hind feet are larger and often show the heel pad, with the rough shape of a human footprint.

Black bears are usually walking, walking, walking, and foraging. Although they do not use the raccoon walk as much as grizzlies or polar bears, I did see that once, very clearly in the soft substrate of a trail I run to the Gila River. Judging from the crisp edges of the track, this was a bear I had just missed.

Like all animals, black bears leave other sign. That's what trackers are really looking for, not merely the tracks of an animal, but *sign*, a tangible mark on the physical world that this animal has been here: walking, trotting,

FIGURE 10.1 Black bear, left hind and left front, black bear, left hind

loping, playing, defecating, feeding, burrowing, scent marking, hunting, killing, dying. Sign has specific meaning—a cottontail rabbit browsed on a woody stem—and larger meaning—we humans are not alone. All these other bodies surround us, suffering and enjoying this same organic life.

Bears typically break the branches of fruit or nut trees, pulling the food closer to eat. They can leave "nests" high in trees that are the debris of this prolonged feeding. Bears bite and claw trees, telephone poles, fence posts, and wooden signs as communication to other animals, including humans. They rub against trees in a seemingly luxurious scratch. They might deliberately chew and break the central branch of a tree as a mark of dominance or territory. They tear up shrubs and trees to build a springy mattress that keeps them off the ground in cold weather. They create worn circular spots on trails that they walk repeatedly, leaving their scent from glands on their feet.

American black bears were once found through all of Canada and the United States and much of Mexico. By the twentieth century, the destruction of their habitat and overhunting had reduced their range to Canada and

FIGURE 10.2 Black bear
(photo by Elroy Limmer)

certain forested parts of the United States. Today, with better protection of animals and land, their population is increasing, with an estimated two hundred thousand in the lower United States, four hundred thousand in Canada, and one hundred thousand in Alaska. Hunters and trappers kill about fifty thousand a year. Black bears can live ten years in the wild and typically weigh from 150 to 500 pounds. In the northern part of their range, these omnivores hibernate for as much as seven months. In the southern part, they may be active all winter.

These bears are generally shy and mild-mannered. Although they are capable of injuring and killing a human being, this doesn't happen often. In North America, wild black bears killed twenty-six people in the first twenty years of the twenty-first century. Females with cubs will bluster, charge, and retreat, playing for time as the cubs climb a tree. The vast majority of fatal attacks actually come from hungry males. Some fatal attacks have occurred when black bears break into isolated homes. Some people have been attacked while running or sitting at their campfires. These are strange, rogue, exceptional events.

For example, about nine thousand black bears live in New Mexico. In 2001, a bear broke into the cabin of ninety-three-year-old Adelia Maestra Trujillo and ate her. Since then, in thousands and thousands of encounters, black bears have run away. We see them from a distance. We see their tracks or scat. We see them up close—and then they run away.

As with mountain lions, the best response to any situation in which you are perceived as prey is not to act like one. Look big and project trouble. Some summers ago, my daughter worked for a natural history museum in Albuquerque taking children on field trips. At the top of a trail in the Sandia Mountains, she heard a low huff, the warning sound of a startled black bear. She did what she had been trained to do. She made sure all the children were behind her and told everyone, calmly, to put on their Big Hat. Putting on your Big Hat means to raise your little seven- and eight-year-old arms above your head, forming a triangle with your shoulders and the tips of your fingers. Maria and her group of children put on their Big Hats. The bear ran away. Nothing more scary than second graders.

My husband and I remember this story fondly. The sky is blue, the air redolent of pine. It's so good for children to be in wild landscapes and see wild animals and know what to do.

We would feel less comfortable if this had been a startled grizzly bear, even though grizzly bear attacks on humans are also rare. Grizzlies are enormously powerful animals, 200–850 pounds, that can live twenty to twenty-five years in the wild. In North America, in the first twenty years of the twenty-first century, grizzly bears killed twenty-nine people. This approximates the number killed by black bears, but there are far fewer grizzly bears than black bears, and they interact less often with humans. Some thirty thousand grizzly bears are in Alaska, twenty-two thousand in Canada, and about two thousand in the lower United States, where they survive in five isolated populations on 6 percent of their historic range. Worldwide, there are an estimated two hundred thousand grizzly bears, mostly in Russia.

Grizzly bears behave differently from black bears, and more than half of those twenty-nine fatal attacks were mothers defending their cubs. While black bears climb trees as defense, grizzlies evolved around fewer trees and don't use this strategy. When you are threatened by a bear who considers you a threat, the best response is not to act like one. Curl up on the ground and protect your neck. This assumes you weren't able to slip away earlier or drive the bear away. Between 2000 and 2015, grizzlies in North America attacked 183 people, and 87 percent of those people survived. So your chances are good. Some fatal attacks by grizzlies were made by startled male bears. One was the result of a mountain biker riding into a bear. Some occurred when hungry bears came on a hunter skinning a deer or moose.

My daughter and the second graders didn't need to think about their best bear response—make a Big Hat or fall to the ground—because the last grizzly in New Mexico was killed in 1931, with a few sightings in the 1950s. But people living with both black and grizzly bears have to be more discerning.

David Mattson worked for many years doing grizzly bear research in the field, in the woods. He never carried a gun and had few confrontations, mainly because he knew enough about bears to avoid them. "It's what you bring to the situation," he says. "It's a matter of feeling centered, being alert." In his work with wild bears, like so many researchers, David marveled at their individuality. Some were overly anxious moms. Some were Mrs. Doubtfire. David and his wife, Louisa Willcox, work now to expand the range of the grizzly bear in northwestern United States, hoping to connect the isolated populations of Montana, Idaho, Wyoming, and Washington. "Don't say wildlife corridor," he explains to me. "That implies something long and narrow, and a grizzly would just claim a stretch of that and block other grizzlies from coming through." Grizzlies need a certain breadth and depth.

David and Louisa talk about the grizzly's "symbolic representation in American culture." For many of us, the grizzly bear is one of our last chances to live in a wild world, which is where we want to live. For others, the grizzly bear is one of our last competitors in the battle for dominance, which is what we want to win. In the contiguous United States, in those places where grizzlies still live, the bears are protected by the Endangered Species Act (ESA). The regulated hunting of grizzly bears is allowed in Alaska, the Yukon, and the Northern Territories. By and large, these are trophy hunts. Grizzly bears can also be legally killed to protect people and domesticated animals.

In 2017, the federal government delisted the grizzly bear as endangered under the ESA. The state of Wyoming, with an estimated 718 bears, immediately planned a hunt that would have killed twenty of them. In 2020, a lawsuit prevented that hunt and overturned the delisting. State management agencies said their desire to hunt grizzlies was based on science and

FIGURE 10.3 Grizzly bear coming out of water
(photo by Elroy Limmer)

conservation: there were too many bears in terms of food supply and potential conflicts with humans. They also argued that old boars, the prize for a sport hunter, were a danger to young bears, particularly females with cubs. As an advocate for trophy hunting wrote, "This is troubling because grizzlies have the second-slowest reproduction rate of any mammal in North America."[1]

When adult male grizzly bears meet females with cubs, they sometimes kill the cubs—and the female if she defends them. This may happen more frequently as a population becomes crowded. This may be how grizzlies self-regulate as they near the capacity of their habitat. In this case, then, hunting is not a scientific intervention so much as interference with a natural process. In 2018, a judge restored protection for the grizzly bear under the Endangered Species Act because the delisting "had not been made on the basis of the best available science." The ruling cited problems with how the number of bears had been determined and the population's lack of genetic diversity.

As with the Endangered Species Act, the North American Model of Wildlife Conservation asserts that science should determine wildlife policy. But some scientists question whether that is happening. In British Columbia, biologists looked at the basis for trophy hunting grizzly bears. They noted that the mortality limits for grizzlies were often exceeded and that there was "considerable uncertainty" in terms of population size, growth rates of populations, and poaching rates. They found a high risk of significant overkill. Moreover, that overkill could easily go undetected.[2]

This wasn't just about grizzlies. The biologists went on to identify four hallmarks of science related to wildlife management: measurable objectives, evidence, transparency, and independent review. They developed a set of "indicator criteria." An indicator of transparency, for example, would be a public explanation for how hunting quotas are being set. The researchers looked for these criteria in hunt management plans across 667 programs in sixty-two U.S. and Canadian agencies, from moose management in Alaska to mule deer management in Alberta. Some 60 percent of those plans met fewer than half the criteria. Only 26 percent had measurable objectives. Less than 10 percent had any form of internal or external review. Only 9 percent explained the technique for setting hunting quotas. These results were published in 2018.[3]

A contrasting model of management is in the territories of the First Nations along the Central Coast of British Columbia, where grizzlies are considered relatives and trophy hunts forbidden. The tribes actively support on-the-ground monitoring of their bear populations, land-use planning that protects key habitat, and projects that promote salmon and other resources for bears.

In terms of trophy hunting, most Americans and Canadians align with the First Nations. Polls in both countries range from 70 to 85 percent against hunting grizzlies for sport. In one American poll, 71 percent marked sport hunting as "morally wrong." These polls were not large, but they ring true. Although numbers change over the years, roughly fifteen to sixteen million Americans have hunting licenses. A much larger number belong to environmental or conservation groups. The Humane Society alone has ten million members. More than 180 million people visit zoos in the United States every year, crowding against fences and barricades, holding up their children, eager for a glimpse of a bear or a mountain lion or a wolf.

David Mattson worries that a small segment of rural America will eventually kill off grizzly bears, legally or not. At the same time, some ranchers and wildlife managers are learning to live with them. They remove animal carcasses that might attract bears. They fence off livestock. They control access to food at campsites and dumpsters. Bears and humans can and do adapt to each other.

David and Louisa want more grizzly bears in Montana, Idaho, Wyoming, and Washington. They also have a "realizable vision" of returning the grizzly to where I live in New Mexico. In his *Grizzly Bears for the Southwest: History and Prospects for Grizzly Bears in Arizona, New Mexico, and Colorado,* David acknowledges that global warming will present challenges, as will opposition from my neighbors here in the Gila Valley and from many Westerners. Even so, we have the habitat, including a Mogollon Complex that starts with the Gila National Forest and the Gila and Aldo Leopold Wildernesses and stretches west into central Arizona. We have, David says, "the moral imperative" to return an animal that lived here for thousands of years before being extirpated by guns and traps in a few decades.[4] And we have the self-interest, the opportunity to be more careful and present in the woods, more physically aware, looking for five-toed tracks and broken branches and marked trees. We have the choice to share the world with an animal larger and more powerful than ourselves because this shared world reflects a larger truth and because we would like, as often as possible, to be awash in that truth.

David Mattson grew up on a small ranch in the Black Hills of South Dakota. The largest wild animals then were white-tailed deer and coyotes. His grandfather had been in the hunting party that killed the last wolf in the state. "As a young teenager," he writes, "I experienced an inchoate sense of loss, largely for reasons I could not articulate, which nonetheless drove me to seek out wild places to work." By the time David was in college, mountains lions and elk had started returning to the family ranch. Today there are black bears. David also cites the extraordinary success of wolves at Yellowstone National Park.

"People with vision," he says, "people with optimism, persistence, and skill can imagine a better world—and make those imaginings come true."

Let me add a caveat. In her book *Wild Souls,* Emma Marris questions the idea of reintroducing species into situations in which they might suffer, either from difficulties with habitat or human society. "Conservation," she writes, "must focus on protecting the ability of ecosystems to adapt and change in a changing world, rather than attempting to stop or reverse all change."[5] Do grizzly bears fit into the culture and ecological future of the Southwest? I really don't know. That remains a subject for informed and thoughtful debate.

Grizzlies are a subspecies of what we call brown bears, which is confusing since the coloring of black bears can also be brown. As well as the grizzly, North America has the Kodiak brown bear, which lives on islands off southeastern Alaska, and the coastal brown bear, which lives along the coasts of Alaska and may be a version of the Kodiak bear. Other subspecies such as the California golden bear and the Mexican bear have gone extinct.

Polar bears are North America's third bear species. They have major symbolic representation. As the planet warms and summer ice melts in the arctic north, the polar bear's prey of ringed seals has become harder to hunt. The pathos of the fallen giant is archetypal. A gaunt figure on a slab of ice surrounded by water. A starving bear scavenging a dumpster before collapsing in the dirty snow. Probably fewer than twenty-six thousand polar bears remain worldwide, mostly in Canada and Alaska, where they can be legally hunted for subsistence and also sport hunted if guided by indigenous Inuit.

Polar bears evolved from brown bears, splitting into their own species perhaps five hundred thousand years ago. While brown bears are omnivorous, polar bears are pure carnivores, with jaws and teeth best suited for chewing blubber. The feet of brown bears have naked soles. Polar bear feet are completely furred and their hairs are hollow inside, another adaptation to extreme cold. Still, as grizzlies expand north because of warmer weather and polar bears move south in search of food, the two species are breeding, their hybrids fertile, with a mix of physical features. Predictably the new bears are called pizzlies or grolars.

Scientists estimate that by 2040 only remnants of ice will remain in northeastern Canada and northern Greenland. By 2050 the population of polar bears will have declined by 30 percent. If arctic ice doesn't stop melting, wild polar bears will be gone by the end of this century.

IUCN BEAR ENDANGERED STATUSES

Black bear, least concern, population trend increasing
Brown bear, least concern, population trend stable
Polar bear, vulnerable, population trend decreasing

Black bears have five toes although their front tracks may not show the smallest Toe 1, which is on the inside of the track. This is the reverse of the human foot, with our biggest toe on the inside. Claws may or may not register. Front tracks can be almost four to eight inches long and slightly more than three to six inches wide. The fused palm pad is wider than long. The circular heel may not register. Fur between the toes and palm pad may blur the track. Since bears carry more weight in their rear, hind tracks are larger. The heel often registers in the hind. These tracks evoke the human footprint.

Grizzly bear tracks can be seven to thirteen and a half inches long and five to almost nine inches wide, bigger and squarer than those of a black bear. Hind tracks are longer and somewhat narrower than front. The claws of a grizzly usually register and are long. There is less fur than in a black bear or polar bear.

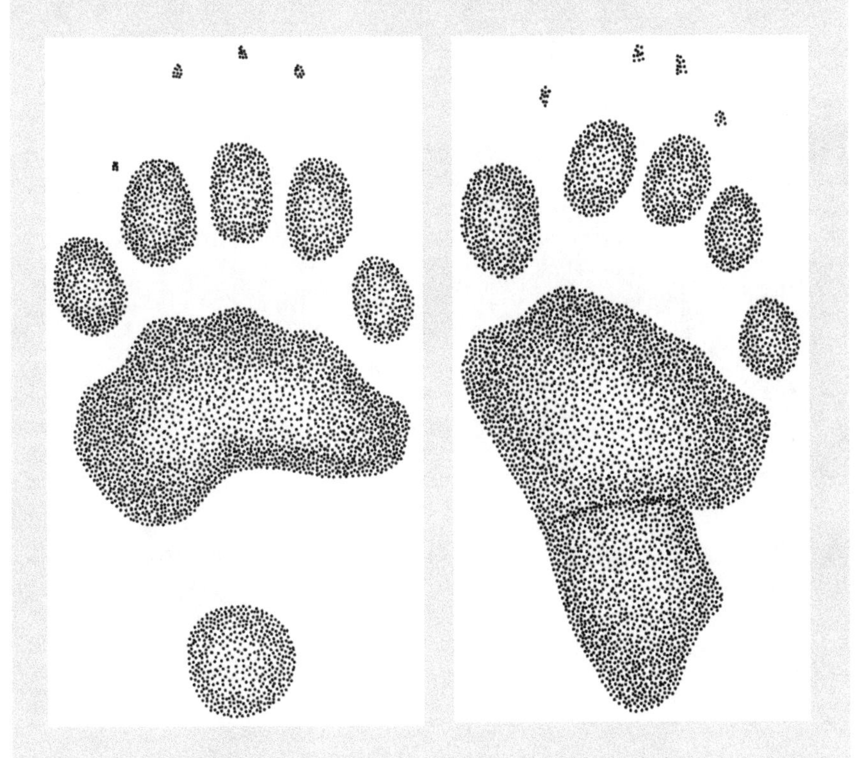

FIGURE 10.4 Black bear, left front and left hind

WILDLIFE TRACK AND SIGN CONVERSATION #124

"Do you ever think about the Pleistocene?"

"I think about the Pleistocene all the time."

"The glyptodont, the four-horned antelope, the ground sloth."

"The three-hundred-pound beaver, the woolly mammoth."

"The short-faced bear, the saber-toothed cat."

"We were outnumbered."

"Outweighed, surrounded."

"That would have been nice."

"That would have been scary."

11

The Sudden Beating of Brains

With any animal track, we want more. That's human nature, wanting more, wanting to follow these tracks, moving through brush and branch, keyed to movement and sound until we reach what the writer Annie Dillard described in her encounter with a weasel: "Our eyes locked, and someone threw away the key."

Annie Dillard's exchange with a weasel "was as if two lovers, or deadly enemies, met unexpectedly on an overgrown path when each had been thinking of something else: a clearing blow to the gut. It was also a bright blow to the brain, or a sudden beating of brains, with all the charge and intimate grate of rubbed balloons."

This is a writer gripped by hyperbole: "It emptied our lungs. It felled the forest, moved the fields, and drained the pond; the world dismantled and tumbled into the black hole of eyes. If you and I looked at each other like that, our skulls would split and drop to our shoulders. But we don't. We keep our skulls."[1]

OK then. Who wouldn't want that sudden beating of brains?

For many of us learning to identify track and sign, however, the etiquette is against it—rather, against foretracking or following an animal's trail. The exception is when the trail is old and there's little chance of meeting the animal. The rule against foretracking can be about the danger of startling a mountain lion or a wolverine. Mostly, it's about not interfering in the life of a weasel or a deer or a deer mouse. These animals need all their energy for everyday survival. No felling of forests or draining of ponds. No grating of balloons. (Except, of course, if you are following that animal to kill it, capture it for research, or take its picture.)

In her essay "Living Like Weasels," Annie Dillard celebrates weasels with figurative language. The weasel she meets is "thin as a curve, a muscled ribbon, brown as fruitwood." He would "have made a good arrowhead." The predator is also a metaphor for following your passion in life, whatever that passion may be. "The thing is to stalk your calling in a certain skilled and supple way, to locate the most tender and live spot and plug into that pulse."

The writer is referring now to the animal's hunting skills. Weasels can hold a rabbit ten times their weight in a full body wrap—all four limbs around prey—as they bite at the nape, trying to sever the spinal column or puncture the braincase. Weasels stalk, run down, lunge, and leap just like the large charismatic predators that we watch on nature shows. They do this all the time, all around us, mostly coming out when the light is dim and we are inside. They go hunting in the burrows of small mammals, exploding the nests of voles and mice and sliding through the snow tunnels of lemmings. Annie Dillard reminds us that this brutal home invasion is "well, and proper, and obedient, and pure." "A weasel," she writes, "doesn't 'attack' anything; a weasel lives as he's meant to, yielding at every moment to the perfect freedom of single necessity."[2]

In this perfect freedom, weasels fall prey to other predators. At two to four months, they leave their mother; most will survive less than a year.

North America has three species of native weasels. Long-tailed weasels are the largest, with males weighing five to sixteen ounces and females somewhat less. Long-tailed weasels range through most of the contiguous United States into southern Canada and most of Mexico. Short-tailed weasels are called ermine but also stoat if you have British in your family. They can be found throughout Alaska, Canada, and parts of northern and eastern United States. Least weasels, with females as small as one ounce, overlap this range and extend further into parts of southeastern United States. When long-tailed weasels have lots of prey and good habitat, there can be fifteen in a square mile. For short-tailed weasels, twenty. For least weasels, sixty-four. The IUCN lists them all as of least concern, with a stable population trend.

The typical gait of a weasel is a bound, a lope, or a mix of bound and lope peculiar to such a long body and short legs. That long body stretches out and then curves in, hind feet landing near or on where the front feet had been. What you imagine is a ribbon in the air. What you imagine is the bright beating of brains and hearts, terror, love, everything.

Figure 11.1 Long-tailed weasel
(photo by Elroy Limmer)

Long-tailed weasels have five toes, although the smallest T1 may not register. The toes look splayed. Claws show usually but not always. These small asymmetrical prints are slightly more than one-half inch to slightly less than one and a half inches in length. Several palm pads can register

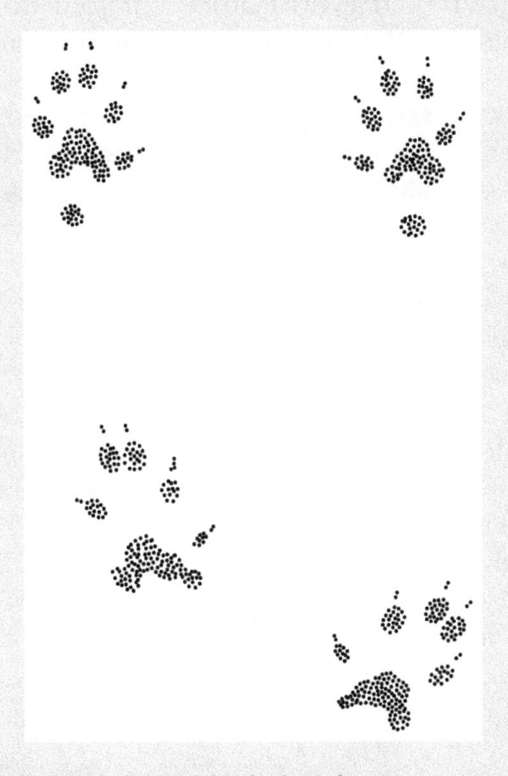

FIGURE 11.2 Long-tailed weasel, hinds above front

separately. A heel pad sometimes registers. The furry feet create a lot of negative space. Hind tracks are slightly smaller than fronts, although they may register as larger, and their two center toes are closer together.

The track of a small long-tailed weasel can look like that of a large short-tailed weasel, while the track of a small short-tailed weasel can look like that of a large least weasel. You know what to do. Get out your field guide.

In the summer of 2019, Sonnie suggested we improve our skills by taking a track and sign class online. We were going out every six weeks as part of the Sky Island Alliance team to document mountain lion, bobcat, and bear, but we didn't do much other tracking and had plateaued at a handful of species. Sonnie explained that the class was on Facebook, a seven-month course offered free by Bob Ollerton, who had a Level Four Track and Sign Certification from CyberTracker North America.

The name CyberTracker is deliberately jarring. In the 1990s, Louis Liebenberg developed a program in South Africa with Kalahari San wildlife trackers. The trackers were given handheld electronic devices with symbols instead of words and a GPS that showed time and location of the animals they found. A software program collected this information, which could be used by park managers and biologists, with the San earning money by doing conservation. The arrangement required accountability—not every San tracker is a great tracker—and a system evolved to evaluate their skills. Certification also opened up other jobs in ecotourism and hunting. Louis wanted the description of his new system to be a little "sticky"—to stick in your head—as well as to evoke both tradition and technology, highlighting that Bushmen were a modern indigenous culture, not relics of a romanticized past.

Thirty years later, the CyberTracker certification is the international standard for determining how good someone is at wildlife tracking. Certifications are awarded after students have been tested in the field and given weighted scores. In the identification of track and sign, a qualified evaluator can give four levels. Level One implies the ability to interpret the sign of medium to large animals, with a fair knowledge of animal behavior. Level Two implies the ability to interpret the sign of small to large animals, including less distinct tracks, with a good knowledge of animal behavior. Level Three implies the ability to interpret the sign of *any* animal, including obscure tracks, with a very good knowledge of animal behavior. Level Four is a score of 100 percent on the evaluation. There are also certifications and levels for trailing or following animals, a more difficult skill.

The CyberTracker process is deeply satisfying to people who enjoy markers of achievement. It's equally warm and welcoming to beginners. The CyberTracker North America website emphasizes that evaluations should be approached as a learning experience. The goal is to promote wildlife tracking. Both for track and sign and for trailing, the two-day field experience will be judged by evaluators who are being paid for their time but not paid very much. The candidate for a Track and Sign Certification will be asked some fifty questions having to do with whatever is found on those two days. After every set of questions, the evaluator explains the correct answer. This is fast-track mentoring. The certifications given—Level One, Two, Three, Four—are just the beginning of something. There are more achievements ahead. Not getting a certification is just the beginning of something. Many people, even wildlife biologists, don't get a certification on their first try.

Bob Ollerton wanted to help people succeed in these evaluations. His co-teacher, Kim A. Cabrera, had the same ambition. Kim has two Track and Sign Specialist Certifications, which are one step above Level Four and given for different regions of the country. Their evaluations have more difficult questions than Track and Sign, and the candidate must get a score of 100 percent. Kim also has a Level Two Certification in Trailing. She is certified in an organization that tracks humans for search and rescue, is a charter member of the International Society of Professional Trackers, and is a curator on the tracking data base for the phone app iNaturalist. She has her own website, www.bear-tracker.com, offering free advice and information. On Facebook she runs a site called Animals Don't Cover Their Tracks. In her busy life as a tracker, Kim is unfailingly patient.

"Studying tracking is a lifetime pursuit. Take your time and enjoy the learning process."

"If you can't identify a track, don't let it get you down."

"You teach yourself by spending time in the dirt."

Bob and Kim had never taught this kind of class before: Sonnie and I would be their beta students. Learning about wildlife tracking on Facebook did, at first, seem jarring. My experience on social media was as a flyby. Taking a Facebook class would mean a regular presence, with Bob posting more than a dozen photos a week that we identified based on material he had also posted. "Who?" Bob would ask about the image of a partial print. "Which foot? Justify your answer."

Bob wanted students who worked or volunteered in wildlife conservation and education. He believed that anyone taking his class should be able to get a CyberTracker Track and Sign Certification within a year. Certainly a Level One. Maybe a Level Two. The first thing Bob did was post a list of mammals common to the San Diego area, where he lived and tracked and where he would eventually get his Specialist Certification and later host an evaluation. Everything in the class would be tied to that list. We wouldn't be studying coatis or black bears. We would be studying coyotes and California ground squirrels and kangaroo rats. Because the fifty questions on a CyberTracker evaluation include the track and sign of all animals, we would also look at common birds, reptiles, and insects in southern California. We would look at skulls and bones. Holes in the ground. Hair on trees. Nibbled grass. Smells. Online? I thought. But I got used to the idea.

Weasels are in the family Mustelidae, which are in the order Carnivora—like every mammal mentioned so far. In North America, the mustelids are weasels, American badgers, river otters, American mink, wolverines, American martens, fishers, sea otters, and black-footed ferrets. Most of these carnivores have short legs, small round ears, and thick fur. The mustelid family is among the most diverse of any. Martens live in trees. Badgers dig holes underground. Least weasels can be as tiny as an ounce. A sea otter might weigh almost a hundred pounds.

In Bob's list of San Diego species, the mustelids were limited to long-tailed weasels and American badgers.

American badgers prefer open grasslands and can be found from southern Canada through most of northern, western, and central United States down into Mexico. The IUCN estimate there are several hundred thousand American badgers in the United States and some tens of thousands in Canada, with two subspecies there recognized as endangered. There are no estimates for Mexico. Although not often seen, badgers are culturally familiar: that striped black-and-white face, the thick muscular body, the long curved claws. A male might reach twenty pounds but can get bigger if food is available. Females rear the juveniles alone until these young disperse, generally between three and five months. Then females, like males, go back to a solitary existence of digging up ground squirrels.

When we do see badgers, they often seem aggressive. Disturbed or threatened, badgers hiss, snarl, growl, and emit a strong odor. They raise their guard hairs to look bigger, stare belligerently, and make mock charges. According to Mark Elbroch and Kurt Rinehart, in their *Behavior of North American Mammals*, "At their highest level of agitation, they also elevate the soft tissues in the nose." This does sound intimidating.

Famously, and more cooperatively, badgers hunt with coyotes. At least, coyotes have been seen guiding badgers to fresh ground squirrel sign by calling or making other signals. The badger digs while the coyote waits for the squirrel to escape from a second burrow hole. The coyote pounces. Sometimes the squirrel runs back instead to the waiting badger. One or the other hunter gets a meal. Sometimes coyotes and badgers hunt with the same partner throughout a season. Sometimes they seem to play or travel together.

We are always surprised when species interact like this. The world so often surprises us. It's true that badgers and coyotes are predators who will kill each other's young and that coyotes occasionally eat badgers. Still, sometimes you make a friend, transactional and work-related.

FIGURE 11.3 American badger with prey
(photo by Elroy Limmer)

The American badger has an impressive track, with five toes, long claws that reliably register, fronts about three to four inches long, and a large, wide, asymmetrical palm pad in the shape of an inverted V. Hind tracks are significantly smaller, with shorter claws. Badgers often walk pigeon-toed. You would be happy all day long if you found such a track, a badger walking pigeon-toed.

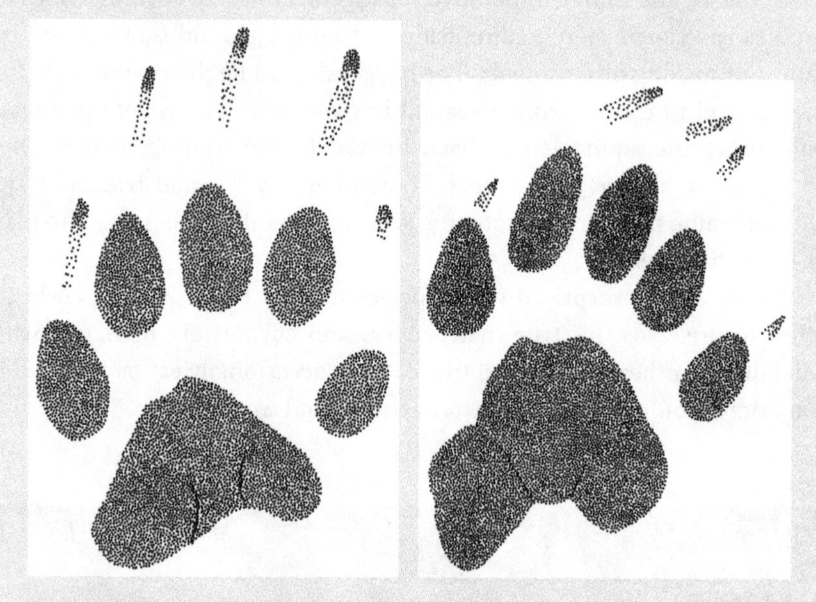

FIGURE 11.4 Badger, left front and left hind track

A third mustelid in the American Southwest is the river otter, which lives along waterways through most of the contiguous United States, Canada, and Alaska. Estimates range from one river otter per mile of coastline in Alaska to one per two and a half miles in Idaho. Defenders of Wildlife believes there are about one hundred thousand river otters in all of North America. This doesn't sound like enough otters to me. But maybe I'm bitter.

For some twenty years, I have been a board member of a grassroots environmental group called the Upper Gila Watershed Alliance, with the unfortunate acronym UGWA. With other environmental groups, we helped fend off a boondoggle dam on the Gila River and the creation of a new Military Operations Area over the airspace of the Gila Wilderness and Gila National Forest. Those efforts took years and were exhausting. Much more fun, we also became part of the New Mexico Friends of River Otters, working to reintroduce that charismatic animal to a state in which the last river otter had been killed in 1953 on the Gila River, near where I live. Finally, in 2006, the New Mexico Game Commission approved a plan to return these fish-eating predators to certain New Mexico watersheds. Between 2008 and 2010, thirty-three otters were flown and driven from other states to be released into a tributary of the Rio Grande River on Taos Pueblo land.

The membership of UGWA, which included many of my friends, had fundraisers for the otters coming here to the Gila. We had films. We had talks. We had T-shirts. We had volunteer days, building the wooden cages where the otters would live as they assimilated on the riverbank before being set free. We imagined them sliding, slipping, swerving, sleek, joyous, all the things we would be if we were densely furred mustelids built for water and play. I remembered my own childhood growing up in apartment buildings in Phoenix, Arizona, where we were lucky to have a small, highly chlorinated swimming pool and a patch of Bermuda grass to shiver on afterward. Marco! Polo! We were ready.

The same people who opposed the Mexican gray wolf also opposed the river otter. In November 2010, New Mexico elected a new governor, who in turn appointed new wildlife commissioners. In 2011, these new commissioners voted to "pause" the release of otters into the Gila River. Allegedly more study was needed to see if otters might affect endangered fish species such as the Gila chub. Groups like UGWA responded that otters would much more likely eat the abundance of nonnative crayfish in the river, who preyed on fish eggs and small fish and were partially why the Gila chub had become endangered. Otters would also prefer the slower nonnatives in the river like carp, suckers, catfish, and bass. Otters would actually be good for the minnow-sized, fast-moving chub, as well as the native loach minnow, spikedace, and Gila topminnow.

Twelve years later, in 2023, otters are still paused on the Gila River. Crayfish and nonnative fish continue to threaten native fish like the Gila chub. Elsewhere, perhaps a hundred otters live and frolic along the Rio Grande in northern New Mexico. Otters have also been successfully reintroduced in nearby Arizona and Colorado.

The river otter has five toes. Sometimes the webbing between toes can be seen. Sometimes the claws register. Front tracks are slightly more than two to slightly more than three inches. The palm pads are strongly lobed. Sometimes the heel registers. Hind tracks are larger than fronts, up to four inches in length, and more asymmetrical, with Toe 1 set farther back and a large palm pad longer on the inside of the print. You'll often find many tracks, since otters are social and commonly in groups. Along a riverbank, look for sliding.

Like many animals in the same family, the tracks of mustelids can be confused with one another. The asymmetrical five-toed tracks of black-footed ferrets may look like the tracks of mink. Martens may look like fishers and mink. Fishers may look like martens and river otters. (River otters can also look like raccoons.) If these mustelids live where you live, and if their range overlaps, you'll enjoy the puzzle of figuring out which is which.

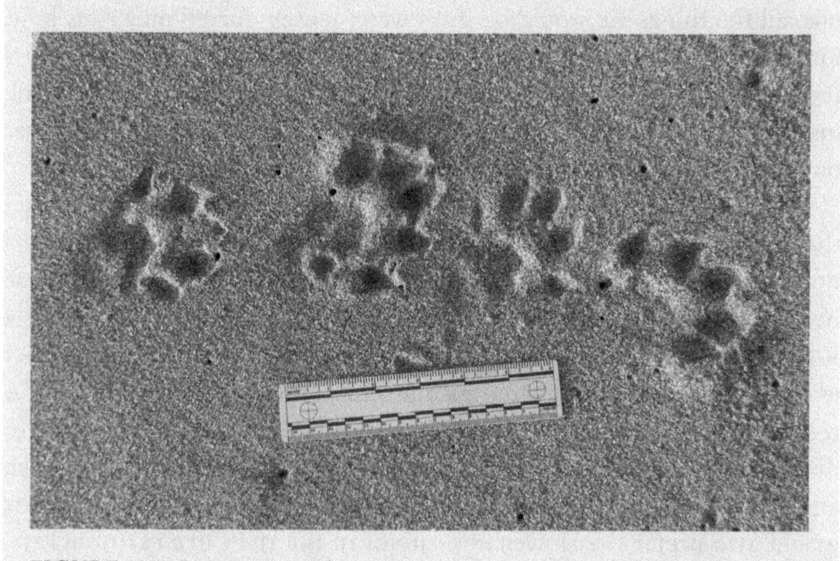

FIGURE 11.5 River otter tracks

The American mink has a range throughout North America, from Alaska and Canada to much of the eastern and central United States. Mink can be found along streams, lakes, swamps, and marshes. Population statistics by the IUCN are brief. The animal is "generally abundant throughout its distribution." In 2005, the density of mink was recorded at one to eight per .38 square mile (one square kilometer). Alternately, in good habitat, there might be nine to twenty-two mink per square mile. That number comes from 1974. We don't seem to know or care how many mink there are in North America, probably because they have become a farmed animal, numbering in the millions. They are also scorned as an exotic species in Great Britain and Ireland.

Wolverines weigh between twenty and fifty-five pounds, the size of a medium dog. Like the badger, they are considered crabby. Like the weasel, they can bring down prey much larger than themselves. According to Elbroch and Rinehart's *Behavior of North American Mammals*, wolverines hunt caribou by dropping from trees onto the animal's back and chewing through the neck. They also slide down snow on their backs, seemingly for fun, and use objects as toys. The IUCN list their populations as the suspiciously precise 3,530 wolverines in British Columbia and "substantial" but unknown numbers elsewhere in Canada and Alaska. There are an estimated three hundred wolverines in the lower United States. In December of 2023, after decades of petitioning by environmentalists, the wolverine was finally listed as a threatened species under the Endangered Species Act.

Martens and fishers live in forested Canada and Alaska and some mountainous areas of western and eastern United States. They have the typical long body, short legs, rounded ears, and thick fur of the mustelid. Both martens and fishers hunt and live in trees, as well as on the ground. Male martens can weigh up to four pounds and male fishers up to eighteen. Females of both species are much smaller. Fishers have a reputation for eating porcupines and house cats. Then again, fishers—also called fisher cats—don't eat fish and are not feline. This may be another misunderstood mustelid. The IUCN believes there are "at least several hundred thousand" martens and "possibly low hundreds of thousands" of fishers.

Two mustelids—the sea otter and the black-footed ferret—are listed by the IUCN as endangered.

Sea otters are a keystone species that maintain kelp forests by preying on kelp-eating sea urchins. Perhaps seventy-five thousand sea otters live along the Alaska, Canada, and California coastline, with more than one hundred thousand worldwide. The threats to these animals include injuries and deaths from commercial fishing, pollution from oil spills, and extreme weather and starvation related to climate change.

The endangered black-footed ferret was actually declared extinct in 1979 because of an infectious plague and human persecution of the prairie dog, its major prey. A few years later, a small population was discovered on a Wyoming ranch. Eighteen wild ferrets were captured, and seven of those became the ancestors of all black-footed ferrets today, about three hundred in captivity and 350 in restoration projects.

IUCN AMERICAN BADGER, RIVER OTTER, MINK, WOLVERINE, AMERICAN MARTEN, AND FISHER ENDANGERED STATUSES

American badger, least concern, population trend decreasing
River otter, least concern, population trend stable
Mink, least concern, population trend stable
Wolverine, least concern, population trend decreasing
American Marten, least concern, population trend decreasing
Fisher, least concern, population trend unknown

WILDLIFE TRACK AND SIGN CONVERSATION #13

"I've never seen a badger."

"I've seen badger burrows."

"Wider than tall, plumes of freshly dug dirt if the badger is active."

"Tracks and sign of the badger's stomach being dragged."

"I watched a nature show once. A group of twenty in someone's yard!"

"Where was this?"

"London."

"Ah, European badgers."

"More urban. More social. More omnivorous."

12

You Think You Know What the IUCN Is Going to Say

Sonnie and I had heartaches, which we talked about when we were first getting to know each other. After that, we didn't talk about them much again. For one thing, they weren't the kind of heartaches anyone could solve. Also, neither of us felt like returning to subjects that we turned over too often in the middle of the night, when the mind experiences a kind of low blood sugar.

Instead, she told me about a black bear she had seen on a camping trip. I told her about the herd of javelina who eat fallen fruit in the orchard near my office window. Recently, she had almost stepped on a rattlesnake. More than once, I have almost stepped on a rattlesnake. We described all the dramatic encounters we had ever had with snakes, including bull snakes that can grow to be so large and impressive. I exclaimed about a YouTube in which a cottontail defended her nest from a large black rat snake, the mother not only chasing away the predator but continuing to pursue the snake up and over a low fence. Sonnie countered with a rabbit she had seen carrying clumps of grass back and forth, back and forth, to line a depression under a bush. The rabbit seemed so intent, not noticing her at all.

We told each other animal stories.

The lagomorphs—limited to one family of rabbits and hares and one family of pikas—are a new order. Carnivorans diverged into cats and dogs, bears and skunks, badgers and weasels. But other animals were evolving, too. King Philip Came Over For Glorious Speciation. Lagomorphs first appeared in Asia and are now almost everywhere.

You think you know what the IUCN is going to say. But the population trend is stable in North America for only one rabbit species and one hare species. The other ten native rabbit species and six hare species are decreasing or unknown. None is increasing. Nine species are of least concern, but one is endangered, one is near threatened, and four are vulnerable.

IUCN COTTONTAIL AND RABBIT ENDANGERED STATUSES

Manzano Mountain cottontail, endangered, population trend unknown
Appalachian cottontail, near threatened, population trend decreasing
New England cottontail, vulnerable, population trend decreasing
Davis Mountains cottontail, vulnerable, population trend decreasing
Eastern cottontail, least concern, population trend unknown
Desert cottontail, least concern, population trend decreasing
Swamp rabbit, least concern, population trend decreasing
Marsh rabbit, least concern, population trend unknown
Mountain cottontail, least concern, population trend decreasing
Pygmy rabbit, least concern, population trend unknown
Brush rabbit, least concern, population trend stable

IUCN JACKRABBIT AND HARE ENDANGERED STATUSES

Black-tailed jackrabbit, least concern, population trend decreasing
White-tailed jackrabbit, least concern, population trend decreasing
White-sided jackrabbit, vulnerable, population trend decreasing
Snowshoe hare, least concern, population trend stable
Arctic hare, least concern, population trend unknown
Alaskan hare, least concern, population trend unknown
Antelope jackrabbit, least concern, population trend unknown

Where I live, desert cottontails and black-tailed jackrabbits are declining because of habitat loss from overgrazing and climate change. Also, federal protection for hawks, owls, and eagles has meant more predators for which rabbits and hares are a staple food. The numbers of coyotes, bobcats, and foxes have increased, too, since the end of widespread poisoning of predators in the 1980s. Importantly, in 2020, a new hemorrhagic virus began to infect wild rabbits and hares, the animals dying of internal bleeding and liver damage. The virus was first seen in the American West and then along the East Coast. It had already decimated domestic rabbits in parts of Europe and China.

"This is the rabbit version of Ebola."[1] Hayley Lanier is a cochair of the Lagomorph Specialist Group at IUCN. Founded in 1978, the specialist group has seventy-three members around the world working on the conservation and management of rabbits, hares, and pikas. Lanier is worried about vulnerable wild native species like the New England cottontail. Domestic rabbits are getting a vaccine.

Even the eastern cottontail, abundant across southern Canada and much of the United States, has subpopulations considered at risk because of predation and habitat loss to agriculture and development. A 2018 study cited by the IUCN has the mortality rate of eastern cottontail populations in some areas as high as 95 percent. That was before the virus.

Still, rabbits breed like rabbits. Eastern cottontails are able to produce a new litter every month for a breeding season that can last nine months. A female might produce thirty-five young a year. Locally, the cycles of estrus, mating, and birthing are orchestrated so that litters are simultaneous, giving predators a shorter time to find hiding places, with a greater chance that some of the kits will survive.

Since the hemorrhagic virus, I still see desert cottontails, although less often. They sit quietly under a mesquite chewing their own cecal pellets, green and soft, straight from the anus. Their short guts, combined with the cellulose in plants, makes coprophagy necessary. Their incisor teeth grow throughout their lives, which means more incessant chewing on grasses, sedges, seeds, fruits, and bark to keep these teeth a reasonable length. All rabbits are strict herbivores. Unlike in the vast majority of mammals, but as in spotted hyenas and baleen whales, female rabbits and hares are larger than males.

As I run by, the cottontail freezes. I imagine Jeff Goldblum from *Jurassic Park* whispering, "Yes, that's Tyrannosaurus rex. Don't move! She can only see you if you move." Then I come too close, and the cottontail does a dash, white underside of tail flashing. This flash will draw a predator's eye away from the body and can produce its own magic trick: the cottontail stops, tucks in the tail, and seems to vanish. Desert cottontails, like a number of other rabbit species, can also climb trees. All rabbits can swim, and two species, the swamp rabbit and the marsh rabbit, do this fairly often.

Sometimes the cottontail doesn't run, unaware or busy, which happened when Sonnie watched that female carry grass for her nest. Cottontails that are molting also use their fur to line nests, actively pulling it away from their body. The kits will be born with eyes closed, covered in fine hair, and able to crawl about only weakly. The mother will nurse them fast in the morning and fast at night, creeping away when she goes out to feed, covering them as best she can, hiding them as best she can, trying not to attract the attention of a weasel or a bobcat or a coyote or a hawk or an eagle or an owl. By two weeks, the kits are moving outside the nest, playing, jumping, hopping. By four weeks, if they have miraculously survived, they are leaving the nest or being pushed out by their mother, who is preparing for her next litter. In some years, such a young rabbit seemed to be crouching under every mesquite bush. They looked stunned, like someone had just pushed them out of a nest.

The desert cottontail has pointy and asymmetrical front tracks, with five furred toes. T3 registers above the other toes. Claws may or may not register. Toe 1 is reduced, clawed, very low in the track, and occasionally registers. Front tracks are about one to one and three-quarters inches long. Hind tracks are significantly larger, up to almost three and a half

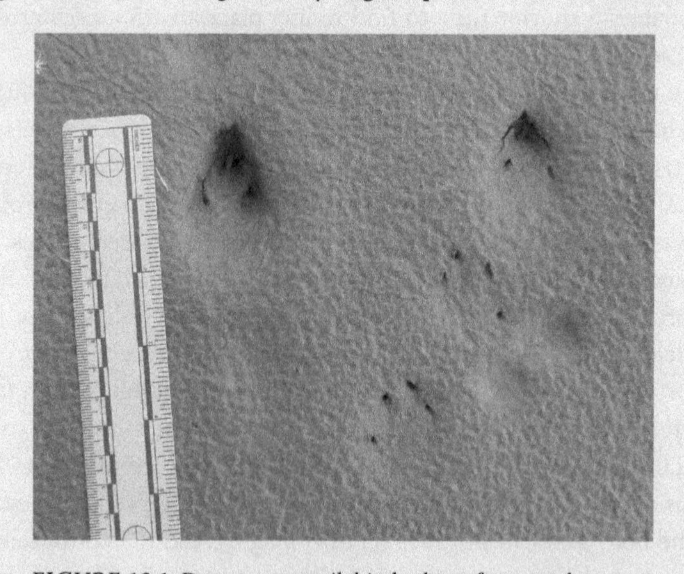

FIGURE 12.1 Desert cottontail, hinds above front tracks

inches, also pointy-looking, with four furred toes that can register and a furred heel. The furry bottoms of these feet often mean an indistinct track.

Hare tracks are larger than cottontail tracks, with a similar appearance. Where I live, a hind track larger than three and a half inches is not a desert cottontail but a black-tailed jackrabbit, which is actually a misnamed hare. The hinds of a black-tailed jackrabbit can be five and a half inches long.

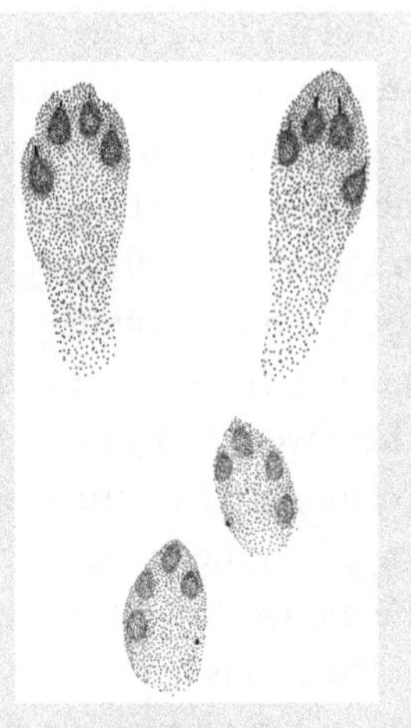

FIGURE 12.2 Desert cottontail in a bound

Rabbits travel in a bound, two hinds above the fronts, and this is what may catch your eye—that pattern in dirt. Hares travel in a modified bound which can resemble a gallop. Sometimes you see the strong claw marks of a "power start." Jackrabbits sometimes jump up in a bound/gallop to look at their surroundings and you can also see this in the track pattern. The distinct pointy look of most rabbit and hare tracks makes them seem like old friends. (Then again, sometimes the hind of a black-tailed jackrabbit will have an impression of claws and toes similar to that of a canine. These tracks can be sly old friends playing a joke.)

Sonnie and I had just finished tracking for Sky Island Alliance, a few hours of finding bear track after bear track in our usual transect in the Burro Mountains. A normal thrilling day following a bear walking up the arroyo. Also, in the mineralized American West, the usual extravagant pink and purple rocks, knee-high, shoulder-high, tumbled together in the streambed. I stopped to admire a boulder with a fist-sized egg of quartz ready to pluck. The largest boulders had that peculiar sentience, murmuring their contentment and prideful age. "Come sit on me," they exhaled. "Come. Sit."

Near our parked cars, in the silky substrate of the dirt road, Sonnie saw the imprint of a black-tailed jackrabbit, which is actually in the genus of hares, not a rabbit. We stopped, got on our knees. This was as exciting as the bear. And this is something to understand about tracks. They can have a value of their own, separate from the animal.

Partly this has to do with the quality of a track. Jackrabbits are easy to distinguish from cottontails if you can see the hinds, which are larger than a cottontail's. These pointed hinds were about five inches long, from the end of my little finger to a certain age spot on my hand, with four claws registering. The hind tracks had swung up in the animal's bound and landed slightly ahead of the two fronts, which were about two inches long, from the tip of my first finger to just past the crease made by the second knuckle. In these fronts, five toes had registered, including the reduced but clawed Toe 1, a mark uniquely far down the track. But then—and what a puzzling thing at first—another pair of fronts could be seen ahead of the hinds. There were two pairs of fronts. That's because when the jackrabbit decided to stop, after its front paws hit the dirt, and as the hind feet landed and stayed on the ground, the animal had put out its front paws again to stop the momentum. And we could see that movement, that stop, right here.

Partly this has to do with time. The black-tailed jackrabbit had been here and in the physical track was somehow still here. The moment seemed to hold both possibilities: here and not here. Particles entangling at a distance. Particles entangling across time. I had a brief insight into quantum mechanics. In fact, I was on the cusp of understanding—and then not. As usual. Farther down the road, we could see more prints of the hare bounding, front feet lifting, hinds pushing off.

The black-tailed jackrabbit is about two feet long and weighs four to eight pounds but seems much bigger with those oversize ears. Unlike

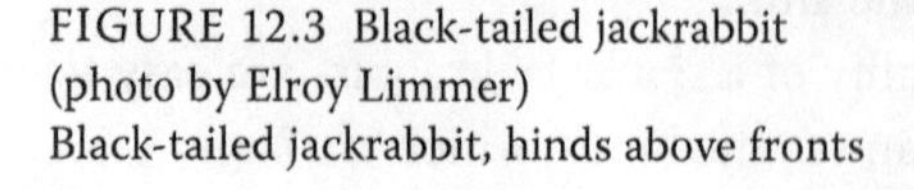

FIGURE 12.3 Black-tailed jackrabbit
(photo by Elroy Limmer)
Black-tailed jackrabbit, hinds above fronts

cottontail kits, leverets are born with open eyes, fully furred, and mobile in minutes. In the day, mothers nurse quickly and then leave their young, who go off individually to hide and survive. At night, she comes back to nurse quickly again and leaves again, worried about attracting predators.

Sonnie and I recognized the track of the black-tailed jackrabbit because we had just covered lagomorphs in our Facebook class. We were on alert for that pointy track, the hind that can resemble a coyote's, and the front a cottontail's. We were looking for prints that had big claw marks, the power start from a sitting position. We often saw blurred prints that were likely jackrabbit because of the gait—what else, that size, bounded like this, showing two hinds above two fronts? We were looking for other sign, too, the way jackrabbits feed on the inner bark of shrubs and differ from rodents by eating deeper into the wood. We looked for their resting spots, depressions or worn areas in the grass or dirt, for the scuffle of a dust bath, almost eighteen inches long and twelve inches wide, and for the round innocuous pellets of scat, about a half inch in diameter, ingested twice.

By this time, in the spring of 2020, the COVID-19 pandemic had begun. Trips were canceled. People stopped having dinner with friends. Sonnie and I still walked outside, looking for track and sign, but six feet apart. Suddenly, a class on Facebook made so much sense.

Some people worry that the American pika will be the first known mammal on Earth to go extinct because of climate change. Adapted to cold climates and high elevation, these small lagomorphs live in mountainous areas of southern Canada and the American West. Scientists have documented the many traits that make them vulnerable. In fact, pikas have disappeared from more than one-third of their habitat in the Great Basin, specifically Oregon and Nevada. Journalists note that pikas, weighing about six ounces each, can overheat and die at temperatures as low as seventy-seven degrees Fahrenheit. Also, pikas do not disperse easily because they cannot move easily into hotter areas. As an alpine species, they are "trapped at the top," one advocacy group warns, vulnerable not only to hotter temperatures but to other effects of global warming—changes in vegetation, extreme weather events, less snow (which pikas use as insulation in the cold), and invading new predators. Although pikas are long-lived for small mammals, they have relatively low reproductive rates. Females usually wean only two young a year.

For much of the twenty-first century, the end has seemed nigh for the American pika. Yet American pikas are not on the U.S. federally endangered list, and the IUCN lists them as of least concern, with a population trend decreasing. Collared pika, in northern Canada and Alaska, are also listed as of least concern, with population trend unknown.

The IUCN has more than 160 specialist groups that focus on species or issues in conservation. Andrew Smith has been studying the American pika for some fifty years and is the IUCN's authority on them in the Lagomorph Specialist Group. In October 2020, he published an essay and literature view in the *Journal of Mammalogy* arguing that the American pika was actually responding to climate change rather well. Smith emphasized that global warming is real and an "existential threat" to the world. It's just not, yet, a threat to pikas.

Although pikas evolved for a cold climate, they have been facing warming temperatures since the Pleistocene. In California, pikas survive at lower and hotter elevations by using talus habitat, or piles of rocks into which the animals can retreat during the heat. Naturally occurring talus is fairly abundant in the American West, although fragmented in its distribution. Pikas are also known to use the tailings left by early miners and the rock skirts that shore up highways. Significantly, pikas can feed on a variety of vegetation and will change their behavior by foraging during the night instead of the day.

Andrew Smith argues that reports of decline are skewed or incomplete. The localized extirpation of pika in Oregon and Nevada, for example, was in their most marginal habitat. Of the 3,250 sites surveyed in the Great Basic, 73 percent were still occupied by healthy populations. Climate changes were about the same in both occupied and unoccupied sites. Many factors can negatively affect pikas, including cattle grazing. As for pikas dying when the temperature reaches seventy-seven degrees, this was the result of an experiment from the 1970s when pikas were left with "fresh green vegetation and a rock to sit on" in an enclosed open cage. Actually, Smith had conducted that experiment. Some pikas did overheat, denied their strategy of retreating into a rock den. Others did not.

The scientist concluded his 2020 paper, "When people learn that I work on pikas, the first thing they say is 'Oh, I am so sorry; you must be sad that they are going extinct.' This information comes from press releases written by journalists after interviewing pika researchers who I believe have overstated their findings. The press releases warn that pikas are being forced off mountain tops; they are not." Moreover, Andrew says, "The discrepancies

FIGURE 12.4 AMERICAN PIKA
(photo by Elroy Limmer)

between available data and their interpretation could be used as evidence that the climate story is manufactured. The credibility of science should be of great concern to all of us."[2]

Today, the narrative of the pika remains interesting for what it tells us about being a scientist or a journalist and the way our understanding of wildlife is fumbling and fraught. Anyone concerned that the climate story will be seen as manufactured, however, can feel somewhat relieved. A few people persist in the fancy that global warming is not caused by human behavior. Many more refuse to take the actions necessary to mitigate global warming. But most everyone accepts the reality of their own experience. The weather is changing and has been changing for years. The days are hotter or colder, more drought or more rain, more worry about wildfires—and then an actual wildfire! More worry about flooding—and then an actual flood! Another power outage. An ice storm. Hurricanes. We are the animal that reads—newspapers, as well as track and sign—and every day we read about the next emergency connected to global warming. The media can't be blamed for these stories. They can barely keep up with them. We may have good reason not to be worried about pikas, but even polar bears are getting less of our collective attention. Mostly, now, we worry about ourselves.

WILDLIFE TRACK AND SIGN CONVERSATION #209

FIGURE 12.5 Ringtail

"People used to see a lot of porcupines in the Gila Valley."

"What happened?'

"I don't know."

"No river otters, no grizzlies. Not many wolves."

"Fewer porcupines."

"I think about the bees."
"I also think about the bees."
"Look at this."
"Maybe . . ."
"Maybe . . ."
"Five toes. A really big pad."
"A big lopsided pad."
"This time you may be right!'
"This time I think I'm right!"
"Ringtail!"

13

The Little Guys

Sonnie had gotten behind in the Facebook class. Trying to catch up, she spiraled into rage. Then despair. All those rodents! I may be exaggerating. But the rodents were very detailed. Bob wanted us to differentiate among the western gray squirrel, the California ground squirrel, the white-tailed antelope squirrel, and Merriam's chipmunk. Also, the black rat, the wood rat, Merriam's kangaroo rat, the Dulzura kangaroo rat, the desert kangaroo rat, the pocket mouse, the deer mouse, and the western harvest mouse. Also, the California vole and the pocket gopher. We should know, too, their scat and signs of feeding and scent marking. Also, their skulls and jaw mandibles.

Clearly this wasn't going to happen.

Even so, the rodents became my favorites. Miniaturization has such strong appeal. A perfect quarter-inch North American deer mouse track is something you'll never forget. You've really tumbled into another world now, everything so much bigger, frightening and exhilarating at the same time.

A mouse track involves some skill. Studying the ground, sometimes on the ground, I contemplate the classic rodent 1-3-1 splay of hind feet, with Toes 1 and 5 out to the side and Toes 2, 3, and 4 connected to the palm pad or not connected to the palm pad, bulbous or slightly bulbous. Or not bulbous at all. The front feet in a rodent have a similar splay but don't actually register Toe 1, which is reduced. Like us, rodents are plantigrade,

FIGURE 13.1 Deer mouse, hinds above fronts

walking on their entire foot, and I think I can see three palm pads and two pads in the heel of the front track and four palm pads and two pads in the heel of the hind track. Possibly I am imagining this.

Size can be important. Kim Cabrera thinks of a wood rat track as a blown-up deer mouse track, with the same bulbous toes not connected to the palm pad. So the question: are these tracks deer mouse size, one-quarter to slightly over one-half inch, or wood rat size, almost one-half to one and a quarter inch? The overlap is problematic.

In other ways, identifying a mouse or a rat can become high-level tracking. More often than not, their prints are only indentations in sand or dirt, without anything resembling toes or palm or heel pads. One trick is to measure trail width, the distance between the track on the left and the track on the right. That distance corresponds to the size and shape of the animal's body. According to Elbroch's *Mammal Tracks and Sign: A Guide to North American Species*, a western harvest mouse in a bound has a trail width of one to one and three-eighths inches, while a deer mouse has a trail width from one and three-eighths to one and three-quarters inches. So if both species live in this area and the trail width is more than one and three-eighths, this is a deer mouse. If the trail width is less than one and three-eighths, this is a harvest mouse.

Unless it's a pocket mouse.

A group of tracks includes all four feet of the animal. The distance between groups of tracks is called a stride, and strides can also indicate what animal made the trail if you have more than one group. A bounding desert wood rat can have a stride up to eight inches, while a black rat can have a stride up to twenty-one inches. A big stride over eight inches would be a black rat.

The gait, too, might be important. Mice and rats usually travel in a bound, hind feet falling above front. Medium to large voles—another small rodent but in another family that includes lemmings—usually travel in a direct-register trot. Of course, mice and rats also walk while foraging for food, and voles sometimes bound.

Just to be annoying, the partial tracks of small birds and frogs can sometimes look like the partial tracks of pocket and jumping mice. Mark Elbroch says, "The secret to avoiding misidentification is to be conscious that such mistakes are not only possible but inevitable."[1]

Wood rats are also known as pack rats, with huge messy nests as big as your house. Not really. But these nests can be more than five feet tall and eight feet in diameter, which seems remarkably large for such small animals. As pack rats urinate on their nests, the crystalized waste helps cement and protect the structure. The same nest might be used by many wood rat generations. In a dry climate, the animal and plant remains fossilize and can be later excavated by human paleoecologists. Radiocarbon dating from these middens have gone back as far as fifty thousand years. This sounds unbelievable. Yet the middens of pack rats are reliable sources of climate change since the time of mastodons and saber-toothed cats. They might also contain odd bits of modern treasure—plastic bags, domestic cow dung, rifle shells, glittering foil, bottle caps, the kinds of things pack rats value even if the rest of us don't.

More than seventy species of rats and mice live in North America, mostly but not all in the family Muridae. Most are doing well, except—as usual—those listed below that live in specialized habitats affected by human development or climate change.

IUCN RATS AND MICE ENDANGERED STATUSES

Salt marsh harvest mouse, endangered, population trend decreasing
Giant kangaroo rat, endangered, population trend decreasing

Beach vole, vulnerable, population trend stable
Texas kangaroo rat, vulnerable, population trend decreasing
Stephen's kangaroo rat, vulnerable, population trend decreasing
Fresno kangaroo rat, vulnerable, population trend decreasing
White-eared pocket mouse, vulnerable, population trend decreasing

Rodents have five toes, although typically only four toes register in the front track since Toe 1 is substantially reduced. Many rodents are diggers or climbers, and their claws show. In the hind track, T2, T3, and T4 are usually close together and point forward. The length of track, the shape and position of toes, and the size, shape, and number of palm and heel pads vary among species.

Deer mice are the most widespread rodent in North America. They move across the landscape in bounds, walking while foraging or under protective cover. Front tracks are about one-quarter to one-half inch, usually registering four toes, which are described as bulbous. Claws sometimes register. Three palm pads often register, with another palm pad and the heel pad registering below these. Hind tracks are only slightly larger, with five toes in the classic rodent splay of 1-3-1. In this pattern, T2, T3, and T4 are close together, pointing up, while T1 and T5 point to the side. Four palm pads often register, with two pads below.

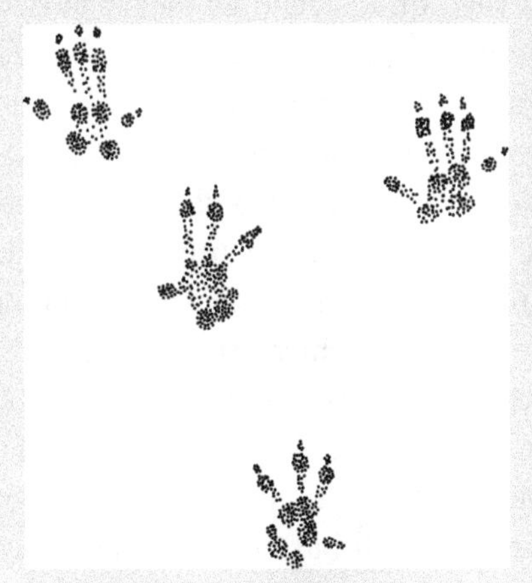

FIGURE 13.2 Harvest mouse

The massive order Rodentia encompasses thirty-four families, 481 genera, and 2,277 species. Rodents (like lagomorphs) have a pair of upper incisors—chisel-shaped teeth at the front of the mouth used for nipping and tearing—and a pair of lower incisors that never stop growing but must be worn away by the self-sharpening act of gnawing. The Latin *rodere* means to gnaw.

Squirrels may be our most visible rodent. In every North American city and suburb, squirrels adorn our parks and backyard bird feeders. Since squirrels do not hibernate, we also see them year-round. Familiarity can breed contempt or, at least, indifference. We see squirrels but don't see them—that bright intelligence and luxuriant tail. We see their faults, how they party in our attic, and not their virtues, how they make a tree come alive with motion, darting at the periphery of vision, reminding us that desire and ambition can exist in so many different forms. We sometimes treat them as pests, an approach almost always doomed to fail. Indeed, the seminal tome *Urban Wildlife Management* says, "Squirrels are so prevalent that any management efforts focused on reducing population is doomed to fail." The authors suggest instead, "Fix the holes in your house." And "Learn to like squirrels."[2]

In the subfamilies of Sciuridae, North America has eight species of tree squirrels, two species of flying squirrels, and fifty-seven species of squirrels more often found on or in the ground—ground or rock squirrels, antelope squirrels, prairie dogs, marmots (including the American groundhog), and chipmunks. These animals vary in size, behavior, and habitat.

Where I live, tree squirrels and rock squirrels take turns at the bird feeder with its solid block of pressed seed on a steel pole meant to deter squirrels. Western gray squirrels seem to have priority, which is both unfair and understandable because these squirrels are so much more glamorous—stars of their own action movie, manic circus performers, leaping and chasing each other in the highest branches of the cottonwood trees. When I don't see them, I hear them. That bark of alarm informs a predator that this gray squirrel is on alert and not worth pursuing. The same small-yapping-dog noise warns other squirrels in the area, too, which are often related.

The counterpart of the western gray squirrel is the eastern gray squirrel, another bold personality and nut collector found in every acre of forest in the eastern United States and much of eastern and southern central Canada. The eastern fox squirrel overlaps much but not all of that range. Both have been introduced as far west as Los Angeles, where they are considered an invasive species.

FIGURE 13.3 Fox squirrel
(photo by Elroy Limmer)

I have come to particularly admire squirrel tracks. Relative to rats and mice, they are larger, with long middle toes that point forward and side toes that can also point forward. In tracker parlance, they have less splay. Claw marks are usually present. Tree squirrels have sharp claws for climbing and hugging trees while ground squirrels that dig burrows have even longer claws. Ground squirrel tracks are more asymmetrical than tree squirrel tracks, with a leading Toe 3 in the front and a distinct curving in of toes that I find satisfying.

Satisfying because pattern recognition is its own reward. Because Bob has explained that one way to distinguish between a western gray squirrel and a California ground squirrel is that leading Toe 3 and distinct curving in, and this information is before me now, confirmed in the physical world.

Ground squirrel tracks can sometimes be confused with those of a spotted skunk. Bob has also explained how to tell the difference. Both tracks have various palm pads and two pads below those, but the spotted skunk track is more symmetrical.

Except for the eight species listed below, all members of the squirrel family are considered by IUCN as of least concern. More than half of their populations have a stable trend, and the rest are decreasing or unknown. Only the urbanite eastern gray squirrel is on the rise.

IUCN SQUIRREL FAMILY ENDANGERED STATUSES

Vancouver Island marmot, critically endangered, population trend decreasing
Palmer's chipmunk, endangered, population trend decreasing
Northern Idaho ground squirrel, endangered, population trend increasing
Nelson's antelope squirrel, endangered, population trend decreasing
Utah prairie dog, endangered, population trend decreasing
Southern ground squirrel, vulnerable, population trend unknown
Townsend's ground squirrel (*U. townsendii*), vulnerable, population trend decreasing
Townsend's ground squirrel (*U. t. nancyae*), vulnerable, population trend decreasing

Western and eastern gray squirrels commonly show four toes in the front and five toes in the hind. When claws register, front claws will be longer than hind. The oval and asymmetrical fronts are slightly more than one inch to almost two inches long, with three palm pads often registering. An additional palm pad and a heel pad sometimes register below. The hinds can be larger, up to three and a half inches long, with T2, T3, and T4 close together and pointed forward. Four palm pads in the hind typically register, and additional pads might register below. These squirrels often travel in bounds, hind feet falling above front feet.

The front tracks of ground squirrels tend to be more asymmetrical than those of tree squirrels, with toes that curve into the centerline of the body. The front tracks of a California ground squirrel are about one to one and a half inches long, usually showing four toes, with Toe 1 reduced. Toe 3 leads the other toes in the front tracks, adding to their asymmetry. The claws are relatively long. Distinct palm pads often show, and heel pads sometimes show. Hind tracks are less asymmetrical, showing five toes in the classic 1-3-1 splay, with T2, T3, and T4 close together and pointed forward. Claws are smaller than on the front track. Distinct palm pads often show, and heel pads sometimes show.

FIGURE 13.4 Western gray squirrel, hinds above fronts

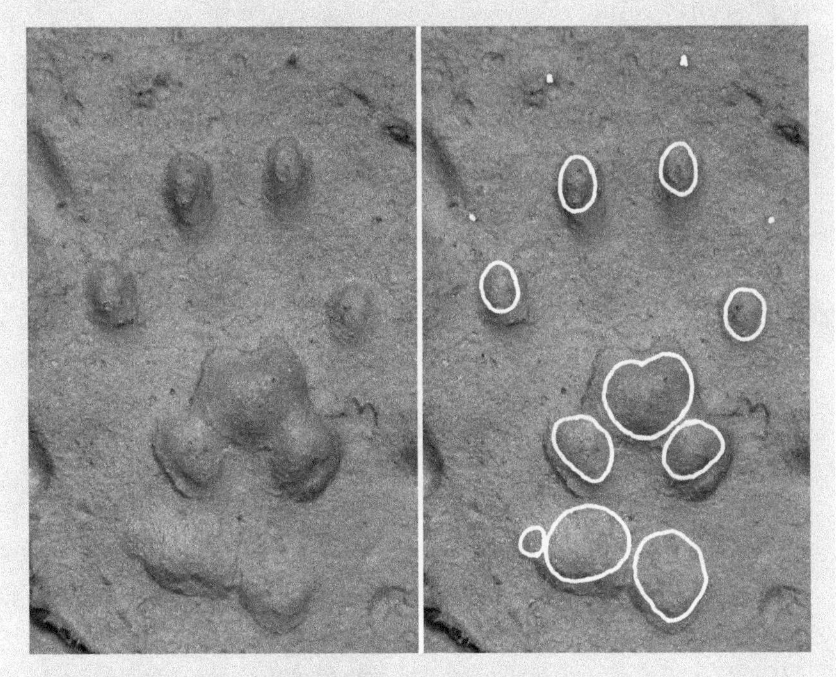

FIGURE 13.5 Western gray squirrel, right front, with front pads outlined

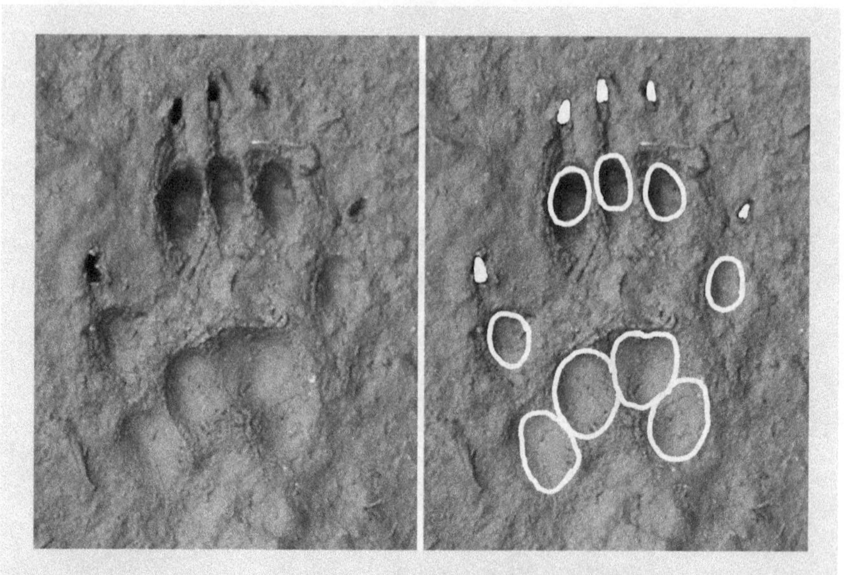

FIGURE 13.6 Western gray squirrel, right hind, with hind pads outlined

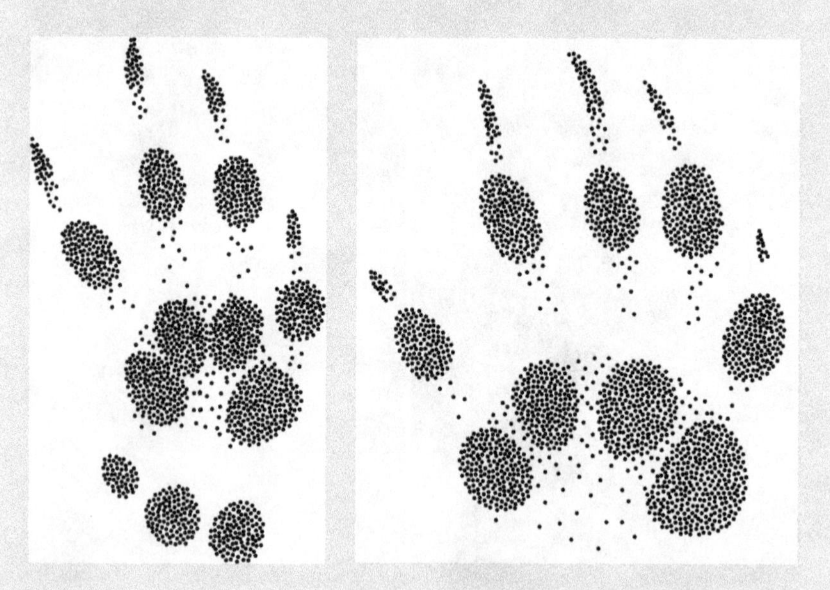

FIGURE 13.7 California ground squirrel, right front track and right hind track

As I was forced to do with mustelids, I am going to descend now into a list—in this case, an incomplete one. There are just too many rodents.

Beavers, for example, are rodents. Although beavers were not on our Facebook list of animals in southern California, Sonnie and I shouted whenever we saw their prints along the Gila River. Beavers have webbed toes, front tracks as large as four inches, and hind tracks as large as seven inches. More often than not, there are only a tail drag and an evocative slide mark into water. Once hunted and trapped nearly to extinction, beavers have returned to most but not all of their historic range, with an estimated ten to fifteen million in the United States and Canada.

North American porcupines are another rodent. The toes of porcupines rarely register, only their claws and a large, odd-shaped palm pad with a pebbly surface. Hind tracks are significantly larger than fronts. North American porcupines can be found throughout most of Canada and the western United States and south into Mexico, as well as in a few eastern states. The IUCN lists them as of least concern, with a stable population that goes through cycles, peaking every twelve to twenty years.

Gophers are rodents. Muskrats are rodents. Shrews and moles are not rodents. Instead of gnawing at plants, these animals catch and eat insects and have sharp, pointed teeth.

Among our native rodents are a few exotics. Black rats were once known as roof or ship rats, famous for bringing bubonic plague to fourteenth-century Europe and now spread through much of the United States and parts of Canada. The larger, more aggressive Norway rat lives in almost all our urban and suburban areas. So do house mice, originally from central Asia. The semiaquatic nutria from South America have invaded much of the American South, as well as the Pacific Northwest and the Atlantic coastline, where they are denuding vegetation in wetland areas, creating open ponds more susceptible to flooding, and outcompeting the native muskrat. We find the tracks of these invasive species and usually feel a lack of enthusiasm. Our hearts do not lift. I am not going to pretend otherwise.

Sonnie called all the mice, rats, voles, gophers, squirrels, moles, and shrews "the little guys." She liked to use the tracking apps on her phone and seemed, in the wild, to be on her phone a lot. Our volunteer work with Sky Island Alliance had us load photos and information into the app iNaturalist. For her camping trips with the New Mexico Wilderness Alliance, Sonnie added to another database called Gila Wildlife, Tracks, Scats and

Sign. Mostly though, in the field, we consulted iTrack Pro, devised by CyberTracker evaluator Jonah Evans.

Back and forth, back and forth—we looked at the small prints of a mouse on Sonnie's phone and the small prints in the mud. The bound pattern, hinds above fronts, was always a pleasure. We agreed to agree that the front tracks seemed slightly larger than the hinds, a deceptive appearance in the harvest mouse rather than a reflection of actual foot size. And Toe 1 in the hinds was connected very low to the palm pad, further evidence of a harvest mouse.

Although Toe 1 in the hind of a deer mouse can also look pretty low to me.

In our Facebook class, ungulates followed rodents. The deer and cows were a relief, so big and relatively easy to identify. The class ended in May 2020. That first year of the pandemic, there would be no CyberTracker evaluation in the field.

WILDLIFE TRACK AND SIGN CONVERSATION #86

"When I was six years old, my older sister said we had a troll living behind our refrigerator."

"That was mean of her."

"No, this was a friendly troll. I was excited."

"I would have been, too."

"One day I went into the kitchen and my sister showed me the troll's footprints on the floor, heading from the back door to the refrigerator. Tiny footprints of mud on linoleum. Now that I remember—this was odd because we lived in Phoenix, in the desert. Hardly anyone ever tracked in mud."

"Yeah, that doesn't sound plausible."

"But the footprints were more exciting than ever. They were proof!"

"Tiny footprints?"

"It was a tiny troll."

"And you have the same feeling now . . ."

"When I see the prints of a pocket mouse. Deer mouse. The little guys."

"What happened to the troll?"

"He disappeared."

"Well, trolls."

"They never write. They never call."

14

Hooves

One day in March 2021, a time when people still hunkered down waiting for their first vaccination, Sonnie and her dog Pumpkin went on a hike with another friend, who is a birder. Birding with Pumpkin was easier than wildlife tracking because Pumpkin naturally became interested in why Sonnie and I were stopping, bending, looking at the ground. "Sit," Sonnie would say, anticipating the problem. "Sit, Pumpkin. No, not there. Not on the track!"

That day, Sonnie and her friend were on a trail I know well in the Gila National Forest. Pumpkin was likely straining at the leash. The birder had binoculars ready. Sonnie, diabetic and with mild emphysema, was happy to be walking slowly, breathing a little hard whenever they went uphill. She turned and said, "It's so beautiful here." I have heard her say that many times, or I have said it first, and she has agreed. Then she fell down, almost immediately unconscious, probably almost immediately gone. That's what our other friend, Ed, said later. Ed was part of the emergency rescue team and the next person on the scene. He was also part of a dinner club that included Sonnie, her husband, my husband, and me, a small group potlucking every few months, exchanging recipes, talking, laughing. Ed thought Sonnie had probably died of a heart attack or an embolism. She was seventy-one years old.

We, all her friends, were shocked. Despite some health problems, Sonnie had been so vigorous. Recently she had taken her first horseback trip into the Gila Wilderness, angling down the steep slopes of those mountains, something she described as exhilarating and frightening. We each had our plans with Sonnie: looking for tracks next Saturday, dinner at a restaurant, her coffee club every week, her work with environmental groups. Sonnie had made our small town of Silver City, New Mexico, thoroughly her home when she moved here some dozen years ago. Now she was gone. Here and not here.

I won't, of course, speak for her husband and daughter. What this meant.

The memorial was held six months later, September 2021. I quoted from Aldo Leopold, "There are people who can live without wild animals and people who cannot." And from Georgia O'Keefe, "I've been absolutely terrified every moment of my life and I've never let it keep me from doing a single thing that I wanted to do."[1]

By then I had started a second Facebook class with Bob Ollerton. Sonnie had talked me into that class, which would end in November with a CyberTracker evaluation. "You and I," she had emailed, "can continue to learn together."

Ungulates—any mammal with hooves—evolved to run faster than the predators that were also evolving to run fast. In two-toed ungulates, Toes 3 and 4 retreated behind a thick horny covering with a hard nail rolled around the edge. Toes 2 and 5 became reduced, higher up on the leg as dewclaws. Toe 1 disappeared. The smaller surface of two toes meant less friction with the ground. Mostly though, with hooves taking the weight of motion, the rest of the foot could elongate and become part of the leg, which increased length of stride. Raising the heels and toes off the ground meant one more joint to push the leg forward, which increased the rate of stride. All this meant more speed.

Hooved tracks are minimalistic. A circle or oval: horse. A heart: deer. Two kidney beans staring at each other: musk ox. Two misshapen tear drops: pronghorn. As relatively large and heavy animals, North American ungulates leave clear and noticeable tracks. Gratifyingly, these often demonstrate gait. You can follow the hoofprints of a deer for long stretches on a dirt road or trail and imagine the animal walking, trotting, loping, galloping, even pronking if this is a mule deer and not a white-tailed deer. When the dewclaws register, you think—ah, moving fast now!

Hooves also show wear and tear. Sharply edged hooves indicate a younger deer, while blunted hooves with chipped edges point to someone older. In many ungulates, males have large antlers or horns, so that front tracks are significantly larger than hinds. Females carry their weight more evenly, with only slightly larger fronts. A mature buck might tend to walk stiff-legged with front tracks slightly outside hind tracks because of the width and weight of the neck and front body, while the doe's wider hips might result in hind feet landing slightly outside the fronts instead. In our romanticized version of a tracker, an attractive man or woman looks up from the ground and says, thoughtfully, "This is a two-year-old doe pregnant with twins who is anxious about finding a place to bed during the day."

Staring thoughtfully at the ground, Sonnie and I would joke about that, too.

"This is a 516-pound female elk who is feeling a little lonely."

"This is a 173-pound male mule deer with a crescent-shaped scar on his right flank."

"This is a juvenile collared peccary who has been playing too many video games."

Where I live in southwestern New Mexico, the ungulates native to North America are white-tailed deer, mule deer, collared peccary, bighorn sheep, elk, and American pronghorn. Farther north, native ungulates include American bison, moose, caribou, musk ox, mountain goat, and thinhorn sheep. These twelve species are all two-toed and divided among four families in the order Artiodactyl.

The nonnative ungulate track I see most often, multiple times a day, every day, is the domestic cow. There are an estimated billion cows in the world and some ninety-four million in the United States, three times more beef cows than dairy. About 70 percent of the western states are grazed by domestic livestock, most of that on public land managed by the Bureau of Land Management and the National Forest Service, as well as national monuments and national parks. In 2021, the U.S. Department of Agriculture reported that 45 percent of the rangeland in the American West was in poor to very poor condition, primarily because of overgrazing by cows and sheep. Commonly, this land shows patches of bare dirt, the grasses are dominated by less edible shrubs or exotic species, and the streams are eroded and drying. There's a reason environmentalists call cows "hooved locusts."

On so many walks, cows, cows, cows.

Cows, cows, cows, and horses, horses, horses—a one-toed ungulate whose middle toe evolved into one large hoof, round in the front and oval

FIGURE 14.1 Mule deer
(photo by Elroy Limmer)

FIGURE 14.2 Deer track with dewclaws

in the hind. Some eight million horses are owned by someone in North America. An estimated eighty-two thousand wild horses range across western states.

Cows, cows, cows, horses, horses, horses, but on every walk, too, on every dirt road and trail—deer, deer, deer. And this makes sense. Deer are prey. For every sign of a mountain lion, I would expect to see many signs of deer.

Mule deer are the more common deer of the American West, extending up into western Canada, with big ears, a patch of white on the rump, and a thin white tail tipped in black. Westerners affectionately call them "muleys" and think of these animals as iconic, like rimrock and certain sunsets. Adapted to dry open habitat, particularly shrubland that provides both browsing and safe cover, mule deer have evolved into a number of subspecies, including the black-tailed deer of the West Coast. Their populations fluctuate based on human behavior.

A good example is Colorado, where gold was discovered in 1859. Anglo prospectors, settlers, and ranchers flooded into the still unrecognized territory, and commercial hunters began slaughtering mule deer to sell for meat. By 1911, a Colorado Game and Fish Commissioner mourned, "The time was in Colorado when deer were so plentiful that it seemed almost impossible for them to be killed off . . . now they must be carefully protected or they will meet the fate of the buffalo and become entirely extinct."[2] Meanwhile, the state's rangeland was given over to sheep and cows, and overgrazing destroyed much of the native grasses. Wildfires were suppressed. Shrublands replaced grasslands. Predators like wolves were killed. Mule deer rebounded. In the 1920s, winter feeding programs began to mitigate the harsh winters. In 1963, hunters had enough mule deer to take more than 147,000. In 1989, the state's population was estimated at six hundred thousand animals, up from some six thousand at the beginning of the century.

In the twenty-first century, mule deer began declining again. Hotter, bigger wildfires burned down some of the shrubland, with invasive plant species like cheatgrass providing less nutrition. Elsewhere, fire suppression had allowed trees to crowd out shrubs. Disease became a problem—a hemorrhagic virus transmitted by biting midges, as well as chronic wasting disease, a contagious neurologic illness. More severe winter weather, related to climate change, meant more deer mortality. A resurgence of elk, which can better survive harsh winters, meant more competition. Migratory mule deer "surf" the green wave of spring growth, and summer drought, also related to climate change, reduced their food sources. Perhaps most importantly, mule deer habitat was being lost to oil and gas drilling and other human development. In particular, the White River herd in northwestern Colorado, often referred to as a "mule deer factory," went from one hundred thousand in the early 1980s to thirty-two thousand in 2013.

The Mule Deer Working Group (MDWG) is made up of representatives from the twenty-four states, territories, and provinces of the United States and Canada that are part of the Western Association of Fish and Wildlife Agencies. Both the MDWG and the private nonprofit National Wildlife Federation identify as pro-hunting. In 2016, the National Wildlife Federation went on record that they considered habitat loss to be the major cause of the decline in Colorado's "mule deer factory." The regional director was clear: "We believe that habitat degradation from energy, and residential development, which has been confirmed by Colorado Parks and Wildlife biologists for years, should be the primary focus of scientifically-based wildlife management."[3]

Even so, in December 2016, Colorado wildlife commissioners unanimously approved a $4.5 million experimental program in which mountain lions and black bears would be killed to see if that helped increase the mule deer population. Scientists, environmentalists, and hunters, too, protested the idea as ineffective and expensive. Local opposition prevented funding until 2017, when the U.S. Fish and Wildlife Service, under President Trump, agreed to take on most of the cost. Groups like the Humane Society, the Center for Biological Diversity, and WildEarth Guardians filed lawsuits. Meanwhile, in northwest Colorado, the program of killing predators began. In April 2021, the U.S. District Court of Colorado ruled against Colorado Parks and Wildlife, saying that the federal government had failed to follow environmental law.

In 2023, winter storms in Colorado dropped more than eighty feet of snow, causing mule deer, elk, and pronghorn to die of cold, exhaustion, and malnutrition. One Colorado Parks and Wildlife manager said, "This year it feels like all we're seeing is starving or dying animals." From another, "My worst days are the days when I have to make the decision to end an animal's life just to end its suffering. It gets to you. The constant calls for sick animals who can't get up along the roadways or in yards."[4] That year, the Colorado Parks and Wildlife recommended an unprecedented reduction in the hunting licenses for mule deer, with licenses for females cut by 94 percent in some areas and licenses for male and either-sex by 48 percent.

Partly in response to concerned hunters, the Mule Deer Working Group produced a 2023 report on mule deer and black-tailed deer in North America. In half their jurisdictions, mule deer were in decline. In the United States, mule deer numbers had dropped to about four million. The Mule Deer Working Group knew these numbers, from Baja California to the Yukon Territory, because deer are so carefully monitored. Like all game animals, they have seasons, they have quotas, they have value.

The decline in mule deer has been accompanied by an abundance of white-tailed deer. This smaller, more forest-loving species has a tail underside that is completely white, so that their rump flashes a disorienting white when they lift their tail and run. In the past few years, Peter and I have begun to see white-tailed deer around our trees and garden, in the irrigated pastures, and jumping across the road. One day we watched a male white-tailed deer guard a doe in the field in front of our house, the two of them nuzzling, his antlers flowing up in a candelabra shape, each new growth coming off the same branch. The next day a male mule deer walked slowly across the same field. This animal was also crowned, antlers forking in a complex pattern arrayed with morning light. Two kings in the field.

White-tailed deer are famous for their comeback. With the arrival of Europeans in North America, commercial hunting reduced some thirty million white-tailed deer to about three hundred thousand. By the early twentieth century, you could grow up in many eastern states without ever seeing a deer. Hunting restrictions helped restore their populations, as did a new landscape of fragmented habitat—suburbs and edges—which these browsers prefer to large expansive forests. Today, white-tailed deer are among our most abundant large mammal. With multiple subspecies, they adapt well to new conditions and are extending their range north into Canada and south to Central and South America. They tend to outcompete mule deer, with whom they can interbreed but usually do not. In the United States, their population is once again about thirty million.

White-tailed deer can live ten years in the wild. Does often birth two fawns a year. Where white-tailed deer have few natural predators, their numbers become a problem. They alter forests by eating all the wildflowers, tree seedlings, and shrubs they like and leaving behind the more browse-tolerant grasses and sedges. They can encourage exotic or introduced species. They eat more plants than there are plants and then they starve. They come into your yard and eat your plants. They cause car accidents. They are vectors for disease. In particular, they have become infected with coronavirus and are now a reservoir, a place for mutations of the virus to develop.

Many ecologists consider white-tailed deer a very big problem. In some areas of the eastern United States, these deer can be fourteen times over the carrying capacity of the land. Oaks and other native trees are not reproducing. Warblers and wood thrushes that live in the understory of forests no longer have brush to conceal their nests. Orchids and trillium are disappearing, along with the insects that depend on them.

In the forests of Wisconsin and Michigan, scientists looked at the impact of white-tailed deer on forest structure. At the turn of the twenty-first century, they resurveyed sites studied in the 1950s and enclosed areas protected from deer today. Professor David Waller, who led the research, concluded that white-tailed deer account for at least 40 percent of the changes seen in these forests over the past half century. He believes the ecological effect of white-tailed deer at "chronically high densities" is not well understood by the general public. Or by decision makers and wildlife managers.[5] Another professor at Cornell University, Bernd Blossey, calls deer "ecological bullies" and exclaims, "The entire food web is unraveling."[6]

The story is familiar. Until gray wolves were reintroduced into Yellowstone National Park, elk had been overgrazing and denuding riparian areas. Aldo Leopold famously described the effect of too many mule deer on Arizona's Kaibab Plateau after he and other game managers had endorsed killing their natural predators. This passage from his essay "Thinking Like a Mountain" is oft quoted and worth quoting again:

> Since then I have lived to see state after state extirpate its wolves. I have watched the face of many a newly wolfless mountain, and seen the south-facing slopes wrinkle with a maze of new deer trails. I have seen every edible bush and seedling browsed, first to anaemic desuetude, and then to death. I have seen every edible tree defoliated to the height of a saddlehorn. Such a mountain looks as if someone had given God a new pruning shears, and forbidden Him all other exercise. In the end the starved bones of the hoped-for deer herd, dead of its own too-much, bleach with the bones of the dead sage, or molder under the high-lined junipers. I now suspect that just as a deer herd lives in mortal fear of its wolves, so does a mountain live in mortal fear of its deer.[7]

When ungulate populations get too big, managers have a few choices, none of them easy. One option is to reintroduce or encourage natural predators. The human bias against other predators often works strongly against that. Another option is to encourage traditional hunting by humans. Notably, however, the trophy hunting of bucks does not reduce deer population, even as fewer and fewer people are interested in the hard work of hunting antlerless deer for food. A third option is to cull animals with hired sharpshooters. This has been successful in some suburban and rural areas. A fourth option is to try to control the deer's fertility through contraceptives or sterilization.

In New York's Staten Island, an area of some half million people, too many deer have been causing traffic accidents and providing a home for ticks that carry Lyme disease. In state and city parks, they eat up native plants, including some endangered species, and encourage the spread of invasive plants. In 2016, Staten Island officials decided to give vasectomies to the town's male deer at a cost of $6.6 million. Six years later, the population of deer had dropped by almost 30 percent, from an estimated 2,053 to an estimated 1,452. Traffic accidents with deer and reports of Lyme disease were also down.[8] In this case, people seemed to want a balance—not an eradication of the deer but a more manageable population that reduced the risks of living with deer. Of course, as deer continue to migrate into Staten Island, more vasectomies will be needed.

White-tailed deer walk on T3 and T4, with T3 slightly smaller than T4. This can help you determine right from left in a single track. The hoofs or cleaves can splay widely. The two smaller ends of the heart-shaped track point in the direction of travel. T1 has disappeared. T2 and T5 are dewclaws on the back of the legs. Dewclaws serve as traction and might register if an animal stops quickly and the lower leg touches the ground. They also register when an animal is running fast or in a substrate like snow. On the front legs, dewclaws show as ovals perpendicular to the track. On the hind legs, where the dewclaws are higher, they show as ovals in line with the track. The front tracks of white-tailed deer are about one and a half to four inches long. Fronts are larger than hinds.

Mule deer tracks are similar, but fronts range from slightly over two inches to four inches. Fronts are larger than hinds. Elbroch notes that

FIGURE 14.3 Deer track in detail

mature male mule deer will have front tracks noticeably larger than females, while white-tailed deer will show less difference in length between male and female. Gait can sometimes distinguish between mule deer and white-tailed deer because mule deer pronk and white-tailed deer do not.

As well as tracks, deer leave plenty of other sign. Mule deer and white-tailed deer browse on trees and shrubs by holding twigs and buds between the hard palette of the upper mouth and incisors in the lower jaw—and then jerking back. This leaves a ragged cut in the remaining plant. Sometimes branches are broken. Deer also leave their scat—familiar pellets. They leave depressions in the grass where they have bedded. They create paths and trails. Males use their antlers to mark trees with physical lines and scrapes, rubbing their scent into the trunk as a visual and olfactory communication to other deer. White-tailed males also scrape the ground with their front feet and urinate in the disturbed dirt.

Hooves have three parts: the hoof wall or hard nail, a more flexible but still hard layer called the subunguinis that connects the hoof wall to the soft pad, and the soft pad. Deer tracks show relatively little of the subunguinis, while elk tracks show a lot. Elk tracks are also larger with wider tips.

An ungulate I see as often as deer is the collared peccary, who come by my office window on their way to fallen fruit from our orchard or a patch of roots or some other errand. Peccaries may have a distant common ancestor with the European pig, but they essentially evolved in North America, the long-nosed and flat-headed peccaries part of that large group of animals that went extinct in the late Pleistocene. When the Central American land bridge formed, some peccaries went south. Today we think of them as pushing up from the tropics, extending their range into northern Mexico and parts of Arizona, New Mexico, and Texas.

In the American Southwest, peccaries are best known by their Spanish name, javelina. About two feet high, weighing between forty and sixty-five pounds, javelinas have relatively large wedge-shaped heads, compact bodies, short legs, and small hooves. Their coarse hair is peppered black and white. Their upper tusks point down, lower tusks up. Their muzzles end in a disk of cartilage. They are mostly herbivores. Of course, they are wary, darting, trotting, jerking out of sight. But javelinas also amble, explore, sniff, become habituated to humans, and seem generally fun-loving.

They come by my office window especially during hunting season. As a protected game species, javelinas can be killed only in certain places and seem to know this doesn't include anywhere near my house. They are capable of breeding year-round, and young males do not necessarily disperse but stay with their families. These groups vary in size. Perhaps a mother, three juveniles, and two nursing babies. Maybe four or five adults and three or four juveniles, and as many babies bouncing around like rubber balls. Maybe a dominant male leads them.

Javelinas might commonly live ten years in the wild. Like humans, elephants, giraffes, wolves, whales—we don't know how many species—they

FIGURE 14.4 Peccaries
(photo by Elroy Limmer)

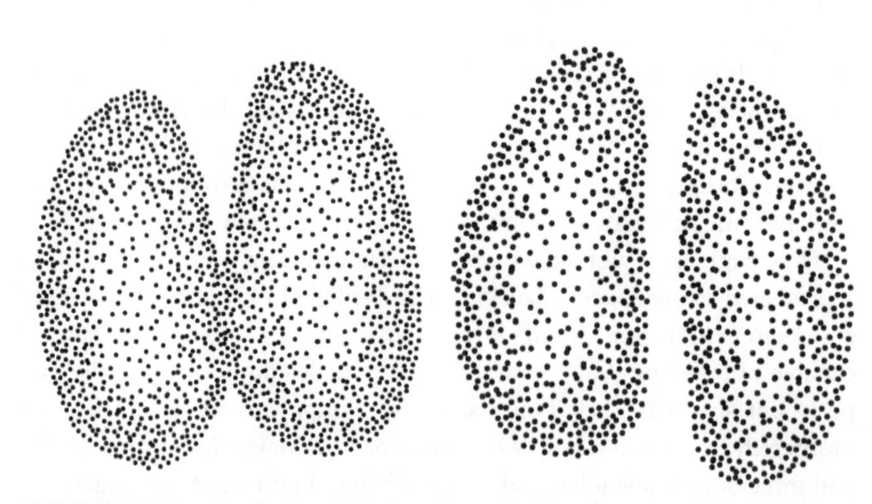

FIGURE 14.5 Peccary, front and hind tracks

show concern at a family member's death. In one case, recorded on camera, a herd visited and spent time with a dead female for more than a week, sleeping next to her, pushing their snouts under her chest, trying to lift her up. One night the group defended the body from a group of coyotes. Later, after the coyotes fed on the corpse, the herd stopped visiting. Javelinas have also been seen lying close to family members who have been caught in traps or shot by hunters, endangering themselves but unwilling to leave.

One day, out walking, I smelled that distinct odor, not quite skunk but equally strong. Javelinas have scent glands on their rump which they use to mark rocks and bushes and each other. A group of eight javelinas froze, half hidden in the grama grass. Although javelinas have poor eyesight, they smelled me, too, and became agitated, milling and clacking their teeth. Some gave out a huffing, woofing sound. Half decided to cross the dirt road to the field. Three jumped the nearby irrigation ditch into a thicket of hackberry and native plum. One began to follow but turned to stare, the hairs on its back erect as it tried to make itself look more intimidating, its scent gland wide open, leaking musk. It woofed. It huffed. Somehow I knew it was an adolescent male.

Then as javelinas do, he panicked and ran in the wrong direction, straight at me, all the while snorting his alarm. The raised hairs gave him a punk look, and he was leaping, covering three feet in a single bound. We were on a collision course until I thought to step aside. He swerved at the same moment. We missed each other by a wide margin. I know that javelina attacks can be dangerous and painful, and I have been nervous myself when faced with a mother and her young. But this guy was too comical. He just made me laugh.

The IUCN gives the population number of collared peccaries as unknown in their native Central and South America. In the southwestern United States, estimates range from one hundred thousand in Texas in 2008 to sixty thousand in Arizona in 2013.

If I felt bad about the terse listing of mustelids and rodents, I feel just as bad about the ungulates. The following species may be ones you particularly love, perhaps that you see often in your yard or outside your window, and I am glad for that, glad for you. In this case, you know their tracks well.

Elk are a large antlered deer that crossed the Bering land bridge into Alaska and lower North America after the glaciers retreated and the mammoths and other megafauna had disappeared. Filling some of those empty niches, they spread across the United States and southern Canada. They remain in scattered populations in Siberia, northern China, Mongolia, and elsewhere; for reasons unknown, they disappeared in Alaska and the Yukon. Their abundance in North America astonished the first Europeans, who hunted some ten million elk to the edge of extinction by the early twentieth century. Two subspecies did go extinct. Today about a million elk live in the American West and parts of southwestern Canada.

American pronghorn also have a stable population of about a million, mostly in the central United States, down from the thirty-five million the European settlers first saw. In another familiar story, pronghorn numbers plummeted to about twenty thousand in 1924 and then responded to hunting regulations and restoration efforts. More closely related to an ancestral giraffe than an ancestral deer, the pronghorn developed its speed—up to sixty miles per hour—alongside the extinct North American cheetah.

When bighorn sheep crossed the Bering land bridge and spread south, they adapted to both cold and hot climates, numbering about two million

FIGURE 14.6 Pronghorn
(photo by Elroy Limmer)

before the Europeans came. By 1936, the Arizona Boy Scouts had started a campaign to save the few animals left in their state. Since then, some eighty thousand bighorn sheep, both the Rocky Mountain and desert subspecies, have been returned to a portion of their historic range. Reintroducing bighorn sheep is not easy. Unlike animals in which the young disperse to new territories, adult bighorns pass on their knowledge of home ranges and migration routes. Transplants have a hard time figuring things out. Wild sheep are also susceptible to infectious diseases carried by domestic sheep. Coyotes and bears are a problem for lambs. Mountain lions are a problem for both adults and lambs. Genetic diversity in small populations is another problem. Finally, bighorn sheep are prized by human hunters, with licenses to hunt them raffled and sold on public and private land.

Of all the ungulates, American bison provide the most famous story of Anglo-American greed and demonic possession. From a high of sixty million in the sixteenth century, covering two-thirds of North America from Alaska into Mexico, wild American bison have plummeted to about thirty-one thousand in fragmented areas of the United States and Canada. This includes twenty thousand plains bison and eleven thousand wood bison; it does not include managed or farmed bison in captive herds.

Moose number about a million animals, living mostly in Canada and the northern United States, dipping into some Rocky Mountain states. In 2015, biologists surveyed the thirty jurisdictions with moose. Population trends were all over—rising, stable, declining, unknown. The basis for these estimates varied. New Hampshire and Vermont used aerial surveys, the observations of hunters, and the number of vehicles that had collided with moose. Other states relied vaguely on "professional opinion." In some places, moose increased as global warming encouraged the spread of shrubs that moose like. Elsewhere, they overheated in warmer temperatures and suffered from blood loss as tick populations exploded. In 2013, Minnesota closed the state to hunting when they realized their northeastern moose had declined by half.

Farther north, caribou have dropped precipitously to some 2.8 million. This number represents a 40 percent loss from previous years in Alaska and a 52 percent loss in Canada. One study points to warmer summers and an increase in insects; the caribou become so intent on escaping their tormentors that they stop eating. The biggest threat, however, is habitat

FIGURE 14.7 Musk ox
(photo by Elroy Limmer)

destruction. Logging and development destroy the forests where caribou live. Mining and dams force them away from migration routes.

Mountain goats live in western Canada, southern Alaska, and parts of the northwestern United States. We assume there are about one hundred thousand. Musk ox are in northern Alaska and far northern Canada—some one hundred twenty-seven thousand. Thinhorn sheep, or Dall sheep, are in Alaska and western Canada—another one hundred thousand. Climate change is affecting all these ungulates. Mountain goats have trouble regulating their body temperature in the summer. Musk ox freeze in extreme storms and starve as rain turns to ice over their foraging plants.

On his website Hunt to Eat: Community, Real Food, and Conservation, Paul Forward describes the decline of the Dall sheep. He begins with a hunt near his home in Eagle River, Alaska.

"In 2017, I drew an archery Dall Sheep tag for the Chugach and spent most of September trying to get into longbow range of an old, mature ram. After 19 nights in my tent and unknown miles and vertical feet traveled on foot, I finally found myself holding the horns of a 9-year-old ram that I had taken with a single arrow at just over 20 yards with my longbow." At this point in his reverie, Paul Forward echoes Dale Lee's admiration for his prey, "He was an incredibly beautiful animal." At the same time, "it was remarkable how little fat he had on him. That hunt remains a lifetime highlight for me because of how much time I was able to spend in close proximity to Chugach sheep. Like that of all Dall Sheep I've had the pleasure of eating, the meat was among the finest food I've ever tasted."

Paul addresses the 50 percent decrease of the area's Dall sheep population. "Why are ewes malnourished to the point of decreased pregnancy and lamb rates? Research is ongoing, but it appears there is a strong correlation between changes in vegetation caused by climate change and the sheep's nutritional status."

In this case, alpine plants were shifting to lower-elevation plants such as alder trees. Also, with the new winter pattern of thaw and freeze, Dall sheep had the same problem as the musk ox, unable to dig through frozen ground to find their winter food. Finally, and poignantly, more winter rain was icing rocks and mountain slopes, and unusual numbers of Dall sheep were falling to their death.

Paul ends his blog post, "I dearly hope that we, the hunting community, can get past political tribalism, accept the science, and tell our representatives that we need an urgent and widespread response to climate change at every level of government. The future of animals that we love and depend on is at the mercy of our decisions."

IUCN UNGULATE ENDANGERED STATUSES

American bison, near threatened, population trend stable
Caribou (reindeer), vulnerable, population trend decreasing
Collared peccary, least concern, population trend stable
Wapiti (elk), least concern, population trend increasing
Moose, least concern, population trend increasing
White-tailed deer, least concern, population trend stable
Pronghorn, least concern, population trend stable
Mule deer, least concern, population trend stable
Mountain goat, least concern, population trend stable
Musk ox, least concern, population trend decreasing
Bighorn sheep, least concern, population trend stable
Thinhorn sheep, least concern, population trend stable

WILDLIFE TRACK AND SIGN CONVERSATION #230

"I guess I'm your tracking buddy now."

"I'm glad. This will be fun."

"This should work out well."

Silence.

"We can go tracking together."

"After all, we've been married for over forty years."

"Right, so this should work."

"It'll be fun."

15

Skulls and Bones

Skulls and bones are part of the CyberTracker evaluation. They are, after all, animal sign. Cow bones are common in the American West. So are the scattered bones of mule and white-tailed deer—that long femur, intricate vertebra, and scatter of ribs. Occasionally there is the full or partial skeleton of a coyote or a gray fox.

In my Facebook class, Bob wanted me to know—which is it? coyote or gray fox? Coyotes have ridges along the top of their braincase that form a V toward the back. So do red foxes. Gray foxes have ridges along the top of their braincase that form a U or lyre, the shape of an ancient Greek instrument. We don't have red foxes where I live. So if the skull has a V, this would be a coyote. A U—a gray fox.

Bob wanted us to identify the skulls of mountain lion, bobcat, badger, northern raccoon, Virginia opossum, striped skunk, spotted skunk, and long-tailed weasel, as well as mule deer, desert cottontail, and black-tailed jackrabbit. CyberTracker evaluations are about what you find in the field in a two-day period, and finding any of these skulls wasn't likely. But the exercise added greatly to our understanding of wild animals and our appreciation of form and function. A predator's canines are used to puncture, hold, and kill. Muscles that attach to the sagittal crest—one of the ridges on the braincase—before stretching to the condyloid process—a protruding structure on the jawbone—give a bite its force. Other muscles attach to prominent

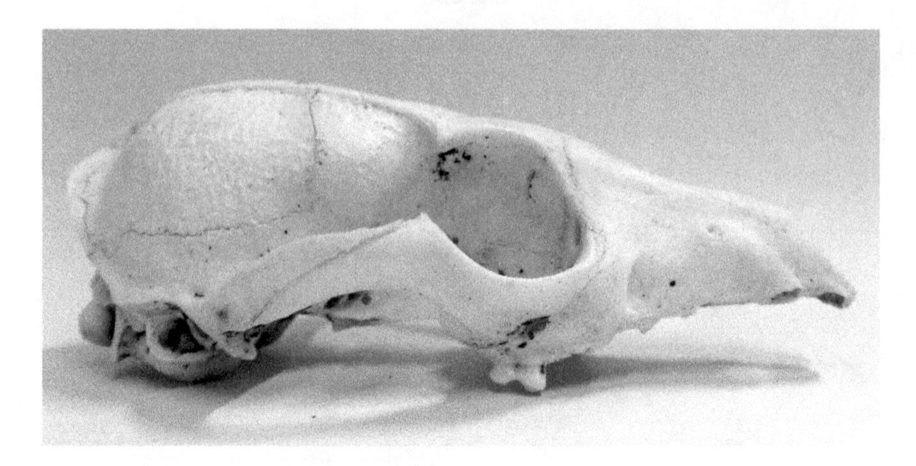

FIGURE 15.1 Gray fox skull

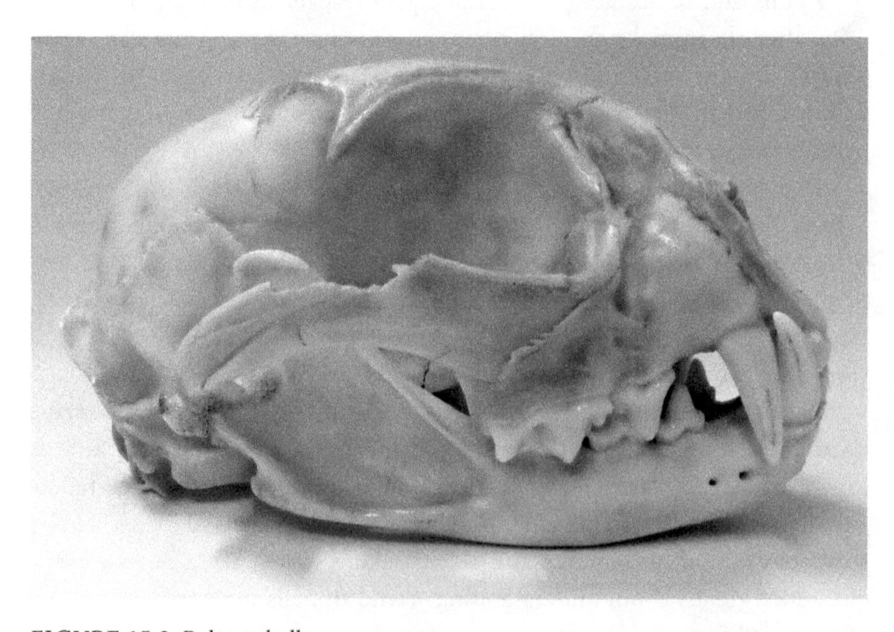

FIGURE 15.2 Bobcat skull

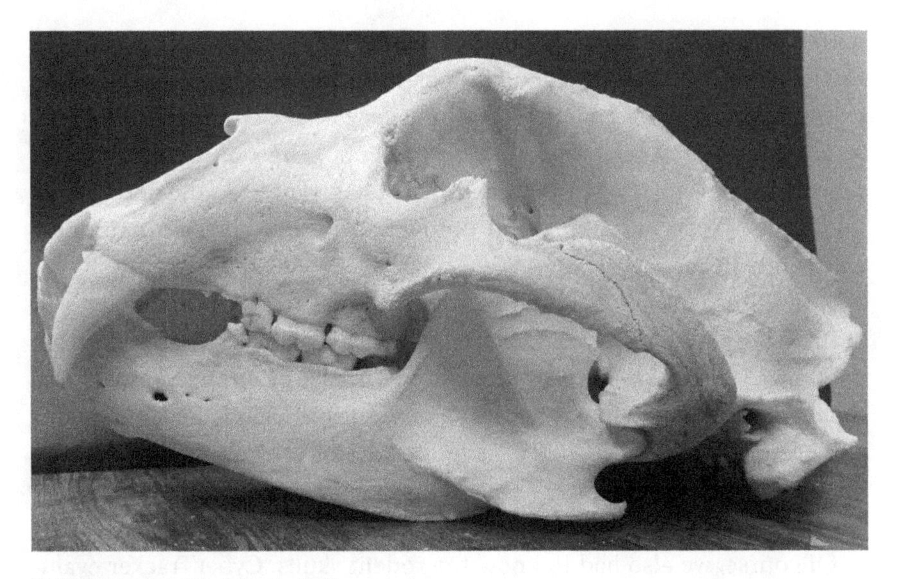

FIGURE 15.3 Black bear skull

points on the braincase so that an animal can chew and gnaw. Eye sockets point forward for better binocular vision to catch prey or outward for better monocular vision to not get caught. A long rostrum—the bones of the nose—increases smell. Large auditory bullae—hollow bony structures enclosing the middle and inner ear—increase hearing.

Learning about skulls is like learning a foreign language. Sagittal crests, occipital crests, condyloid processes, auditory bullae. As Mark Elbroch points out in his *Animal Skulls: A Guide to North American Species*, common names do not exist for many parts of the skull. This new vocabulary "requires a certain enthusiasm and even stamina."[1] (If you have *Animal Skulls*, as I do, filled with page after page of Elbroch's elegant and meticulous illustrations, you fully understand how much this man represents genius in the world of wildlife tracking. Also how easily science crosses the boundary of art.)

Before this Facebook class, in decades of hiking the American Southwest, I had never been tempted to take home the few animal skulls I found—skulls still attached to bleached bones or solitary like an abandoned ball. That kind of ownership, a collection of skulls, seemed both macabre and disrespectful. Now something shifted. These artifacts were not only useful to me but also deserved my attention and praise. I began researching other cultures that value skulls and bones—certain tribes in North America, that Catholic church in Portugal—and returned again to Georgia O'Keefe and her famous imagery of skeletal remains against lucent blue, often accompanied by flowers. "The bones seem to cut sharply to the center of something that is keenly alive on the desert even though it is vast and empty and untouchable—and knows no kindness with all its beauty."[2]

And knows no kindness with all its beauty. I think of Sonnie, "It's so beautiful," and then she was gone. In many ways, skulls are different from the usual treasures I have brought home, rocks of attractive color and shape, photographs of land and sky, memories of clouds and mountains that seem eternal. The skulls of animals are not eternal—but neither are they frail. They exist somewhere in a liminal space, emanating the power of living and dying, alive and mortal. Here and not here.

Of course, we also had to know the rodent skulls. CyberTracker evaluations do find owl pellets, the regurgitation of a bird that has swallowed its prey whole, and within those pellets will be the indigestible bones of a small mammal. Bob highlighted eight rodents.

A Botta's pocket gopher skull is the most common skull found in an owl pellet in southern California, so I paid special attention to this animal's robust cheekbones, or zygomatic arches. The cheekbones form an opening that allows muscles and tendons to pass underneath, connecting the cranium to the jaw. Pocket gophers need a lot of muscle to break down their preferred food of tough woody plants, and their wide arches—wider than the braincase itself—indicate that a lot of muscles are passing through this opening. Carnivores and many omnivores also have wide arches, but ungulates chewing grass and shrubs do not.

Another common skull is the kangaroo rat, easily distinguished from the pocket gopher. Kangaroo rats have spectacular hearing thanks to their auditory bullae, or hollow bony structures around the ear, that create chambers that resonate with the near-silent wing beats of an owl or the quiet movement of a snake.

Unfortunately, the mandibles, or jawbones, of the pocket gopher and the kangaroo rat are sometimes the only remnants of the animal that you find, and they can be easily confused. Both mandibles have deep notches and a lateral extension. Both have four peglike, rootless, basinlike cheek teeth. But! As Bob explained in regard to the kangaroo rat, "The first cheek tooth is not hour-glass shaped as is the case for pocket gophers."

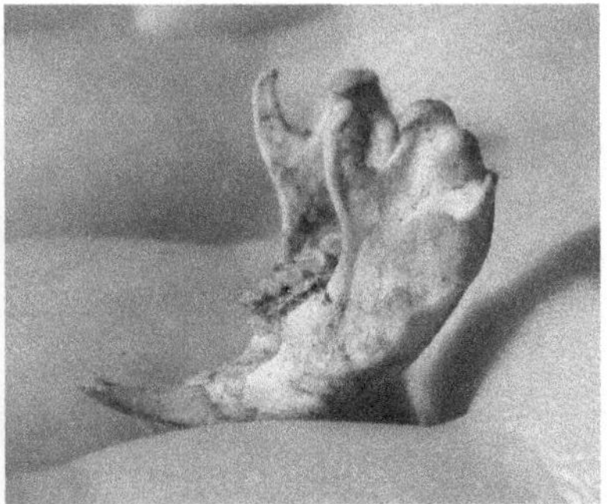

FIGURE 15.4 Pocket gopher skull and mandible

And so it goes. The angle of teeth in a California vole slants downward toward the back of the skull. The braincase of a woodrat is round. The braincase of a black rat is also round, but the cheek teeth are rooted and moundlike, not basinlike. The skulls of deer mice and pocket mice are smaller than those of woodrats and black rats. The California ground squirrel has postorbital processes—bone extensions that support and protect the eye—and temporal ridges that form a V where they join the sagittal crest.

As with anything, such details become more obvious and meaningful when you give them a name. Then, when you start looking closely at where wild animals live, you begin to find a surprising number of bones. Often deer bones. Sometimes a coyote. But also very small bones. The remains of a vole or a mouse or a rat. You might find many small bones in a pile, with the partial skulls of many species. Out of this jumble you pick up a mandible that fits on the end of your finger, its prow curving up like a Viking ship. You think "exquisite," a word you had never before associated with a kill site.

WILDLIFE TRACK AND SIGN CONVERSATION #300

"Raccoon walk! Right front, left hind."

"Here's another."

"Later, a squirrel coming by."

"Skunk. Seam in the track. Striped skunk."

"Heron."

"Kill deer."

"I love these panels of mud along the river."
"They're like paintings."
"Time capsules."
"Lizard? Tail drag?"
"Woodrat? Tail drag?"
"Lizard or woodrat."
"I think so, too."

16

A New Vision of Wildlife Management

In September 2021, the Southwest Environmental Center launched its national campaign, Wildlife for All. The goal is to reform wildlife management in the United States. Its executive director, Kevin Bixby, says, "Wildlife management in every state is stuck in the past."[1] Wildlife for All wants systemic change. They want legislation that protects nonhunted species (their preferred term for nongame animals), including reptiles, birds, and invertebrates. This would reflect our modern understanding of ecology. We live on a planet in which all the parts work together. We need pollinators like bees just as we need predators like mountain lions.

They want a legal acknowledgment that wildlife and healthy habitat have their own value apart from what is useful or valuable to humans. This does not mean prioritizing animals over humans. No one is going crazy here. This is about being better humans, deepening our understanding of ethics and responsibility.

They want wildlife agencies that reflect the broad public interest in wildlife, not just the special interests of a few groups. People who identify as hunters, trappers, fishers, farmers, or ranchers should be represented but not disproportionately.

They want dedicated funding sources for wildlife not tied to the sale of hunting and fishing licenses or to taxes on guns, bullets, and archery equipment. Federal programs like State Wildlife Grants and the Recovering

America's Wildlife Act can provide the money needed to protect non-hunted species of concern. Portions of a general sales taxes or property tax can go to wildlife conservation. We fund highways. We fund libraries. We can fund wildlife.

If you are an American, Wildlife for All suggests you learn how wildlife decisions are made in your state. Their website has the feature "Find Your State," which will tell you about that. Start going to wildlife commission meetings and other meetings related to wildlife management. You'll learn a lot (I did), and other people may learn from you. Since the governor often appoints wildlife commissioners, talk to your governor, too. In your state legislature, identify wildlife champions and support them. "Don't get discouraged," Wildlife for All warns, "if your efforts are not immediately successful."

If you are a Canadian, do similar work in Canada.

Wildlife for All envisions a new North American Model of Wildlife Conservation in which "all living beings are treated with compassion, live in coexistence with humans, and are respected for their intrinsic right to exist and for their role in natural systems."

Compassion. Coexistence. Intrinsic. I've been steeped in such language for more than fifty years now. In 1975, I was a college student getting a degree in Conservation and Natural Resources when I first heard the term *deep ecology*, coined by the Norwegian philosopher Arne Naess. All life on Earth has the right to survive and flourish. We are all connected. We all have value. Naess had been influenced by Aldo Leopold's *A Sand County Almanac*, published in 1949, and Rachel Carson's *The Silent Spring*, published in 1963. In turn, the deep ecology movement of the twentieth century rested on millennia of human thought from cultures and peoples around the world. Humans are not separate from nature. We are part of something larger. Ever since I was a child, I have recognized these ideas as true, true for me, bedrock true, true in my bones.

When I asked Kevin Bixby if he had read these authors, he began to quote from them.

What I want to say is that we still have a lot of wildlife in North America, and we should celebrate these animals. We don't see them much. I live next to three million acres of national forest, and I don't see them much. But we can be alert to their track and sign. This can become a form of seeing.

I want to exhort: we must ally ourselves with wildlife. We must work to mitigate global warming, keep our public lands ecologically healthy, alter landscapes for humans in ways that consider wildlife, and reform our management of wildlife. We must choose wildlife. We must demand more wildlife, more of the nonhuman world, more celebration, more humility, more empathy, more connection.

I want to say again that this is not a call to end hunting or trapping, commercial or recreational fishing, or the killing of animals that threaten people and, to some extent, property. This is not a claim to know what the future of our coexistence with wildlife will look like. Respecting the rights of nonhumans to flourish is complicated and fraught. This cannot be overstated. Squirrels get in our attic. The skunk sprays our dog. Grizzly bears make us nervous. To live with wildlife is to enter an entangled world that is "uneasy, asymmetrical, and tenuous." I might add mysterious and paradoxical.

I know some things: no more killing contests, trophy hunting, leghold traps, and any human behavior that causes unnecessary suffering. Also, indigenous hunting and gathering cultures and people who see wildlife as relations have a lot to teach the rest of us.

I have focused in this book on the track and sign of mammals, and I have left out so many of them. The opossum leaves a striking track that Bob Ollerton says "resembles the marks a tarantula might make falling from the sky." This is partly because the animal's hind foot often lands on the highly splayed front track, which has been compared to a court jester's hat. Armadillos have distinct long middle toes. Bats leave the sign of tubular droppings.

There are more than 457 mammal species in northern North America. Close to half again as many reptiles and amphibians and probably twice as many birds. Some ninety thousand species of described insects, with perhaps as many undescribed. There are field guides to the tracks of these animals, too: *Tracks and Sign of Reptiles and Amphibians: A Guide to North American Species* by Filip Tkacyk; *Tracks and Sign of Insects and Other Invertebrates: A Guide to North American Species* by Charley Eiseman and Noah Charney; *Bird Tracks and Sign: A Guide to North American Species* by Mark Elbroch and Eleanor Marks. I have a growing collection on my shelf. They make me happy just looking at them.

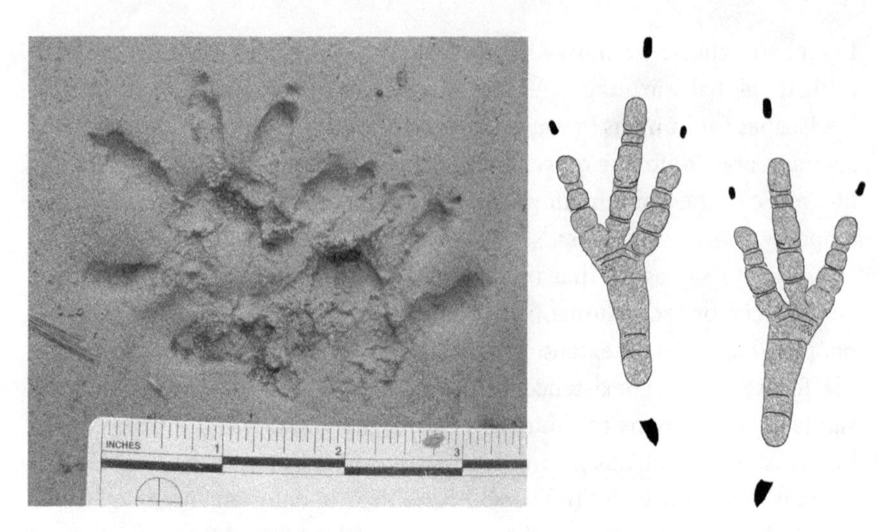

FIGURE 16.1 Opossum and raven

WILDLIFE TRACK AND SIGN CONVERSATION #22

"I didn't bring my reading glasses."

"Next time."

"And the tracking book."

"Which one?"

"The heavy one."

"Do you think this is a porcupine?'

"No."

"Let's try the phone app."

"Next time I'll bring my reading glasses."

17

The Evaluation

An hour's drive east of San Diego, I stood with an earnest group of people around a large scat that had been circled and marked with a red surveyor's flag. On that November morning at Sycamore Canyon Open Space Preserve, we wore sweaters and coats we would take off as the day warmed. From this dirt parking lot, the view was undulating creosote and bursage turned green with recent rain. Low hills in the distance erupted with rocks reminiscent of teeth and fists. The sky was the usual infinity and a promising shade of blue.

Jonah Evans gave a short speech, welcoming us to the moment. Today he would be the evaluator of eight candidates, with Bob Ollerton and my husband, Peter, as his assistants. Just ten minutes before, Jonah had surveyed the dirt parking lot and found three animal signs for us to identify. A popsicle stick with the word Who? pointed to a pile of scat, a second popsicle stick to a track nearby, and a third popsicle stick to a second track.

The procedure was to look at whatever Jonah circled and popsickled, write down your response, and then quietly approach one of the recorders, Bob or Peter, and show them what you had written. Throughout the day, Jonah and Bob went ahead to find more track and sign, which they circled in the dirt and left with a new question. Usually there were about five circles, each marked by a red flag. Some of the popsicle sticks had more than one question: Who? Which foot? Gait? The eight candidates went from circle to circle, decided on their responses, had them recorded, and then—when everyone had taken all the time they needed—Jonah explained the correct answers.

The scat was horse: large, dried, biscuity. One of the tracks was also horse. The second track was that of a bird. The corvid pattern of Toe 2 close to Toe 3 fit both crow and raven, but a raven's tracks are larger than a crow's. If this track was more than three and a half inches, the answer would be raven. I had estimated the distance from the tip of my finger to the base of my finger as about three inches, but I hadn't practiced much with three and a half inches. I felt shy about getting my finger too close to the ground and smudging the truck. Actually, I felt shy about putting my finger into the circle at all, which had become a kind of sacred space. I also worried that I was taking too long, that another candidate waited impatiently behind me. This wasn't true but felt true. Moreover, we have ravens in the Gila Valley, not crows, ravens as ordinary as coffee in the morning. I was here in exotic California. I put down crow.

Jonah gave the answers. Horse, horse, raven. I had missed one already.

Let it go, I thought. That would become the day's mantra. The inner spiritual work. I wasn't here to get answers right. I was here to learn. Let. It. Go. Then Jonah explained that the first three questions were just for fun and practice. They would not be part of the fifty questions on the evaluation. Oh, OK! I let it go.

Later I would notice that the more experienced members of the group got quite close to a circled track and decisively used their fingers as rulers, still careful not to touch the ground. The most experienced member made a drawing of every track or sign marked by a popsicle stick, no matter how obvious the answer might seem. He bent down. He lay down. He took all the time he needed.

Our evaluator was *the* Jonah Evans, developer of the phone app that Sonnie and I had used so often in the field. Jonah had grown up on a ranch in the Hill Country of Texas with parents who started an education center for wildlife conservation. After working in wildlife education himself, he took a nine-month apprenticeship with trackers Mark Elbroch and Jon Young. That was the start of something. Eventually he became a CyberTracker Track and Sign Specialist and a CyberTracker evaluator, along with working full-time as the state mammologist for Texas State and Park Wildlife. As an evaluator, Jonah's style is relaxed and reassuring. He has a soft voice. He smiles easily. For this week in southern California, my evaluation was his third, back-to-back. I thought he looked a little tired.

Another CyberTracker rock star in our group was Jonathan Poppele, author of numerous animal track guides, a Level Three from Minnesota taking this evaluation as a way to explore a new area with new species. Two other candidates had gotten Level Three in previous evaluations and hoped now to get Level Four, which requires a score of 100 percent. Above all, the three men said, you learn a lot. One woman had taken the evaluation before but not reached any level. Four of us, all women, were taking the evaluation for the first time. One of them was my friend Alison, another good friend of Sonnie's, who had also been in Bob's Facebook class.

Sonnie would have appreciated Jonah's gentle opening trick of easy questions. Horse scat. Horse track. Raven or crow? A way of helping beginners spread their wings. Sonnie had a big laugh, and I could almost hear her boom beside me. She whispered, "So what did you put down?" It was okay to whisper now, after the answers had been revealed. "Crow," I admitted. "Darn," she said.

The eight candidates, plus Peter, waited while Jonah and Bob inspected the dirt road that led up to this parking lot for more track and sign they could use for the evaluation. Dirt roads so often have good substrate along their edges.

On this dirt road, I got mule deer right and the gait of walk, a perfect set of tracks spread out for more than twenty feet. I got coyote right, the Who? apparent in the size of the track, compact four toes, negative space showing X, and relatively small palm pad. I knew it was a left foot, being on the left side of the trail, and a hind foot, being above the front foot in what looked like a straddle trot, a common gait for a coyote. I got wood rat nest right, that messy pile of sticks. I got bobcat right and felt the usual thrill: clearly feline, C in the negative space, teardrop-shaped toes, big pad. I *knew* this track, a right front in a walk. I got dog right. Big happy-go-lucky tracks.

Another thrill now—the small bipedal hop of a kangaroo rat. I had seen these prints so often on Sacaton Mesa in southern New Mexico. I remembered the young mountain lion hunter and his pack of hound dogs. And if these were kangaroo rat tracks, this little hole here marked with a popsicle stick was likely the digging of a kangaroo rat, not a burrow and seemingly too deep for a dust bath. But some exploration. And this scat, too, right next to these miniature tracks and hole, with its own popsicle stick. More kangaroo rat.

I got that one wrong. I should have recognized cottontail scat, so much larger and rounder. In fact, I had had a niggling concern—wasn't this scat so much larger and rounder?—before faulty correlation swept that aside.

Cottontail scat is such a common sign that CyberTracker probably rated this question as easy. Missing an easy question is not good. In the weighted scoring of an evaluation, the candidate gets one point for the correct answer to an easy question or three bad marks for a mistake. Two points are given for the correct answer to a difficult question or two bad marks for a mistake. Three points are given for a very difficult question or one bad mark for a mistake. Your sum of points for correct answers is divided by your sum of points and bad marks, and that percentage is the final score. A score of 69–79 percent is Level One, a score of 80–89 percent is Level Two, a score of 90–99 percent is Level Three, and a score of 100 percent is Level Four.

By the end of these two days, there would be eight questions on scat, and I would get three of them wrong. That was one lesson: learn more about scat. Another lesson was something Jonah would cite casually—a study in which experts on scat were right 95 percent of the time in the field and 50 percent when given the scat in a plastic bag. The scat of too many species can look like the scat of other species—gray fox confused with javelina or raccoon with black bear. What's important is the location of scat and your knowledge about an animal's behavior. Foxes like to leave their scat on an elevated surface. Bobcat scat will often have a scrape.

After our group of candidates answered the questions posed at the Sycamore Canyon Open Space Preserve, we got in our cars and drove to the Scripps Poway Parkway Wildlife Tunnel, also inland of San Diego, also someplace Peter and I would never have found without Google Maps. Sonnie rode with us. "Really?" she said. "You missed cottontail scat?"

But I was feeling good. At the wildlife tunnel, I recognized the Who? of a partial raccoon track—although not Which Foot? failing to remember that Toe 1 of the hind sits so much farther back than the other toes, something that distinguishes hinds from fronts even if not all the pad has registered. I recognized mule deer again and knew this was a front track because of its larger size than the hind. I correctly wrote down cottontail and identified wood rat scat with an associated nest.

Then I mistook a fox track for coyote. Once again I mismeasured, going too fast. The size alone was enough of a clue. I looked at the track of a spotted skunk and wrote down California ground squirrel, front left, in a lope. The last two answers were right. But Bob had given us explicit details about the difference between California ground squirrel and spotted skunk. And I didn't not remember those details! Only . . . hadn't someone said spotted skunks were rare in this area? They are rare where I live, too. Spotted skunks are like double rainbows. And the tunnel had been so dim. Poorly lighted. The track had really seemed to curve inward like the track of a ground squirrel. I couldn't really make out the palm pads. Maybe, again, I had gone too fast.

Only the recorders knew what answers we gave them. As Jonah talked about the tracks, most of us kept a poker face. We were adults. We didn't need to punch our fists in the air or suddenly sob. It's true that Bob came over afterward and asked if his information about California ground squirrels and spotted skunks had been deficient. Oh no, I said. Not at all. This was entirely about me. My preconceptions. "It's what you always told us, Bob. Don't let your preconceptions get in the way. Look at what you are actually seeing. This is a great example of that." I was as embarrassed as I had been in a long time. Oddly, too, and I knew this was odd, I felt I had let these animals down. Gray fox. Spotted skunk. They deserved better.

We got back in our cars and drove a half hour now to the Magnolia Avenue Bridge in another part of the San Diego megapolis. Bob had warned us that tracking in any urban area meant going under freeways or bridges for which we might want flashlights and earmuffs for the noise. The Scripps Poway Parkway Wildlife Tunnel had been dry and isolated from human presence. Under the highway bridge, though, pools of water and slick rock dominated, with sodden and discarded shirts and underwear, bags of fast food, the debris of people living here or traveling through. A depressing place, I thought, still rattled about the spotted skunk.

But two men came by walking their dog and were perfectly cheerful. "What are you doing?" they asked. And "What have you found?" Jonah gave them a tour of the circles and red flags.

In the next few hours, I would identify the tracks of another cottontail and a striped skunk, a hummingbird nest, a cliff swallow nest, lizard scat, a California ground squirrel burrow, and trails left behind by beetle larvae. I missed Which foot? of the striped skunk. I confused the track of an opossum with another California ground squirrel. I thought aged domestic dog scat was aged bobcat scat. And I misidentified a bobcat track as a raccoon. Mark Elbroch's *Mammal Tracks and Sign* would comfort me later: "When Toe 1 is absent, the rear track [of a raccoon] is often mistaken for a bobcat track." Jonah himself said, "This is a bobcat pretending to be a raccoon pretending to be a bobcat."

Mostly I was comforted by dinner that night in Little Italy, San Diego, and a half bottle of wine. Over clam linguini, Peter could afford to be philosophical. "Let it go," he said. As a recorder, he was learning a lot.

Before leaving the Magnolia Avenue Bridge—forever, I hoped—I found myself walking with Jonah back to our cars. He talked briefly about his wife and two children. I asked how many evaluations he did a year, and he said more than six started to feel like too many. Of course, he said, he always learned a lot. You learned something every time you went out into the field. Sometimes he fantasized about leaving his full-time job just to do wildlife tracking. But he liked his job, too, he said quickly. It was all good. Being outside. Being with good people.

We talked next about climate change. Everyone does. The website for CyberTracker warns, "The climate crisis and the potential extinction of up to a million animal and plant species may pose an existential threat to humanity." Monitoring that threat, in terms of the health and abundance of wildlife, can involve tracking skills. In an article for the *Journal of Wildlife Management*, for example, Jonah analyzed how well biologists were surveying river otters in Texas. Twenty-three field staff in the Texas Parks and Wildlife Department had been part of a seven-year study that looked at otter track and sign. When they took the CyberTracker evaluation, however, they misidentified the tracks of twelve other species as otter. The three most otterlike tracks are swamp rabbit, raccoon, and opossum. The most experienced observers, who had been with the survey more than two years, misidentified eleven of forty-two tracks as one of these species. They also misidentified otter tracks as nonotters.[1]

Jonah concluded that many wildlife studies needed a method of teaching and assessing tracking skills. The evaluation itself was educational. The scores of these biologists improved dramatically in a second evaluation three months later.

The next day of my CyberTracker evaluation was another hour's drive, this time from San Diego to Lake Moreno. More Sturm und Drang of freeways and cars as suburbs and box stores became mountains and grassland. Another meeting of our parked vehicles and exchange of hellos. Then the impact of landscape seemed to hit us all at once. The quiet and dreaminess. Green-leafed cottonwood, majestic oak, yellow grass, blue sky. Not much moving. We exhaled collectively.

Throughout the morning, we would notice the absence of the eighth candidate, who had gotten lost trying to find Lake Moreno. Periodically, too, Alison and I exchanged glances and shrugged philosophically. We had made so many mistakes. We had no hope of getting certification for any level. Sonnie had disappeared for most of yesterday after I misidentified the aged dog scat but reappeared now, beaming good humor. "It's so beautiful," she said. And it was so beautiful, midmorning, with those green trees, yellow grass, and blue sky. We were outside doing something interesting with good people. That's something I never take for granted. In November 2021, after the deaths of some eight hundred thousand Americans from COVID-19, the deaths of millions of people worldwide, not one of us took that for granted.

For the first scat, people started taking out their magnifying glasses. The men who had already reached Level Three were especially intrigued. I gave a guess—a gray fox eating fruit and seeds—but this turned out to be wild turkey. The big crumbly brown scat was unusual for its absence of white uric acid, since birds don't produce urine but excrete their nitrogenous waste as a cap of white paste. At least, they usually do.

"When anyone asks me, 'Does it ever?'" Jonah said, "I always say yes."

We walked on to look at and identify a gopher mound. A bobcat scat and scrape. The ridgeline of a mole. Jonah circled the tracks of quail, scrub jay, and California thrasher. Here a deer had rubbed his antlers against a tree, the stripped bark obvious, and the elegant mark of tines. Here a toad had walked across the dusty road, with a scat—unexpectedly large—in the middle of the tracks. After the answers, Jonathan Poppele laughed, "And did you see the drag mark? Like some kind of tail? I was trying to figure that out and then realized he was just dragging his scat behind him as he walked."

This was such a personal view of animals. We noted where a cottontail had sat down, where a roadrunner crossed the trail, where another roadrunner had left his scat some time ago—ropy, globby, with lots of white uric acid. We were in their world, practically in their digestion system. The dreamlike landscape was becoming more familiar and less dreamlike, full of a life and activity I had not seen before.

I was also thinking: how could anyone mistake fox scat for turkey scat?

I felt bad about the deer tracks, too. The question wasn't Who? But What Gait? The distance between tracks looked too short to be a lope or gallop. It didn't resemble a trot or walk. I should have calculated the length from shoulder to hip of a typical deer, estimated length of legs, and visualized how deer legs move. I should have drawn a diagram. But as with those dreaded word problems in high school, I had been seized by a sudden blankness of mind—an encounter with the void. I guessed walk. The answer was lope.

For their last animal track and sign, Jonah and Bob led us farther down the road, another half mile. We walked and talked and thought about lunch, having tracked for some hours. No doubt, a few of us still parsed out our errors. But all that fell away in the excitement of the badger burrow, a dark inviting hole about nine inches wide, in sandy soil. The tracks at the entrance were blurred and old, although we all looked for the five toes, long claws, and asymmetrical pad. After we gave our answers to the recorder—every one of us getting this right—Bob said he had seen the burrow active a few years ago. He had actually seen three badgers in the distance. Likely they were a mother and juveniles. Today, looking into the distance, I could almost see them, too.

Back at the cars, we each ate the lunch we had brought. I envied one man's oily peanut butter and gleaming jelly on slices of home-baked wheat bread. Someone's deviled eggs also looked good. Peter and I had perfectly adequate crackers, cheese, and olives from a deli in Little Italy. Jonah sat by himself and figured up the scores.

The math behind weighted scores is complex. But the evaluator only needs to add up the points that the candidate got for each right answer—one, two, or three points depending on the difficulty of the question—and the marks against for each wrong answer—one, two, or three marks depending on the easiness of the question. In most cases, the questions fit into well-known categories, and their ratings of difficult to easy have already been determined by a CyberTracker committee. Next the evaluator does a simple division, the total number of points by the sum of points and bad marks.

After lunch, Jonah gave another speech: how well we had done, how impressed he was, how much fun he had had. He didn't sound insincere. Then, one by one, he handed out the certificates.

Level One to Allison. She blinked in surprise and pleasure. It's the first time I have actually seen that happen, except as a cliché in a book.

Level Two to me and two other women in the group. I blinked in surprise and pleasure.

Level Three, with quite high scores, to the three men who already had Level Threes.

No one made a 100 percent that day, so no one achieved Level Four. Notably, in 2022, only twelve people in North America had a Level Four in Track and Sign. Many more people had Levels One, Two, and Three, all thanks to the work of teachers like Mark Elbroch, Casey McFarland, Jonah Evans, Bob Ollerton, Kim A. Cabrera, Jonathan Poppele, Janice Przybyl, Bryon Lichtenhan, Rosemary Schiano, and others.

My certificate, which read Track and Sign II, had the illustration of a spotted leopard looking alert and a San bushman about to shoot an arrow. Jonah had signed his name above the acknowledgment "According to standards set by Louis Liebenberg and CyberTracker Conservation." Next we were presented a brown and green badge, embroidered with another San bushman, which theoretically could be sewn onto a jacket or shirt.

This was the beginning of something. For some of the people here, this was a ladder to keep climbing. They hoped to get a better score on the next evaluation, a Level Two, a Level Three, a Level Four, and then a 100 percent in a Specialist evaluation—as Bob had done recently. They might start studying and taking Trailing evaluations. Achieving some level in both Track and Sign *and* in Trailing would give them the actual title of Tracker in the CyberTracker hierarchy. Maybe they would get a Level Four in Track and Sign and a Level Four in Trailing and become a Senior Tracker, of which, in 2023, there were only ten in North America. Maybe the level beyond that: Master Tracker.

For all of us, this was the beginning of being more alive and aware of the animals and plants around us, learning to slow down, bend down, look, and really see. Love, that psychic substratum, the measure of every good day's end. By lunchtime, this had already been a good day. The memories that remain are the prints of a toad in dust and the lines of an antler on a cottonwood tree. A bobcat scrape. Turkey scat. Yellow grass and blue sky.

Epilogue

An early reader of this book suggested I not include the chapter you just read. She thought some of the mistakes I made might undermine my authority as a tracker and your guide to this introduction to track and sign. I half agreed. But I also felt stubborn. Because mistakes and confusion are part of wildlife tracking. The mystery of an animal here and not here. The embarrassment of your ignorance. Your assumptions. Letting that go.

As Kim assures us, "Studying tracking is a lifetime pursuit. Take your time and enjoy the process."

After my first evaluation, I decided not to pursue more levels in the CyberTracker world. I had achieved my goal. In the fall of 2023, after also taking the Facebook class with Bob, Peter received a Level Two certificate along the sandy stretches and bosque of the Rio Grande, under a highway bridge, the trucks and cars of Albuquerque passing steadily overhead. He saw beaver, badger, porcupine. Lots of raccoons. Striped skunk. Deer mice. Bobcat. Gray fox. Bats. Snowy egret, Canada geese, roadrunner. Cicadas. Bagworms. At the end of the day, the evaluator Casey McFarland sat cross-legged before a half circle of trackers also sitting on the ground. Casey brought his hands together and cupped them at the level of his chest. He suggested that the trackers do a mental exercise. He wanted them to remember everything they had witnessed, the track and sign of so many

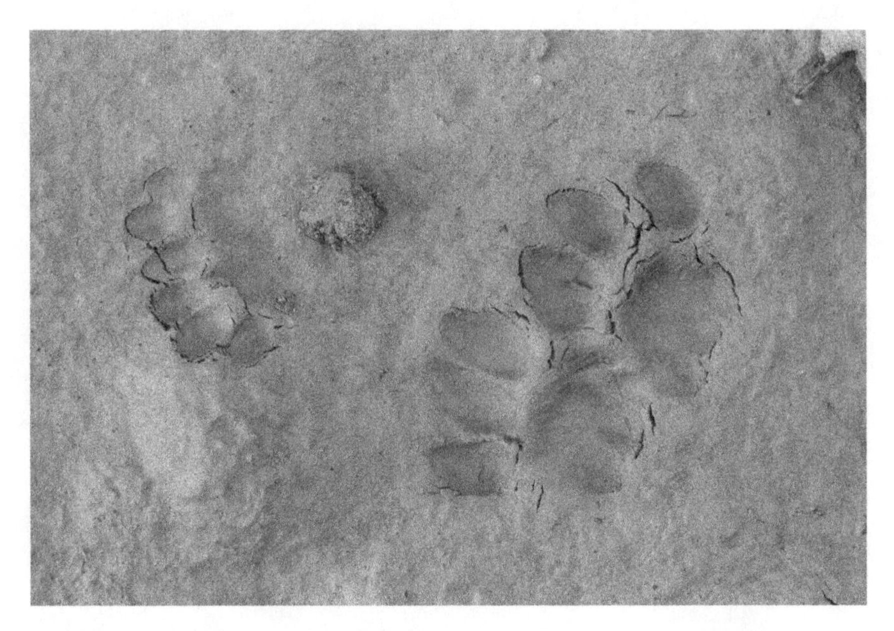

FIGURE E.1 Ringtail, double register, and Kim's cat Beez

animals, and bring all that—this very ecosystem—into their bodies, their rib cages, their hearts.

Well, maybe I'll take another evaluation with Casey McFarland. You learn a lot.

Meanwhile, my life as a tracker continues with my husband and other friends along the trails, dirt roads, and arroyos near my home, along the banks of the Gila River, that green ribbon of cottonwood and willow, brown hills flapping out like wings. In memory of Sonnie, I look for the catlike track of the ringtail. Sometimes, I feel the presence of Coyote. Delightful existence.

Acknowledgments

I would like to thank all the wildlife trackers mentioned in this book and many other trackers whose works I have read or who have accompanied me in the field or in a workshop. Special thanks go to Harley Shaw, a good editor and friend. And to Elroy Limmer, who takes such evocative photographs of the world and generously shares them. Except for photos attributed to Elroy, all images—photographs and illustrations—are by Kim A. Cabrera, expert tracker and wonderful collaborator.

Many thanks to Miranda Martin, who supported this book so early in the process. Also thanks to production editors Michael Haskell and Ben Kolstad, and to copyeditor Peggy Tropp. Tremendous thanks to Peter Riva, who never fails in his support. Finally, always, love to my husband, Peter, my lifelong tracking partner.

Notes

1. RIBBON OF LIFE

1. Aldo Leopold, *A Sand County Almanac* (New York: Random House, 1966), xvii.
2. World Wildlife Fund, *Living Planet Report 2022*, https://www.worldwildlife.org/pages/living-planet-report-2022. Also see Yinon M. Bar-On, "The Biomass Distribution on Earth," *Proceedings of the National Academy of Science* 115, no. 25 (2018): 6506–11. Among the earlier scientific works to promote the term *defaunation* is Rodolfo Dirzo et al., "Defaunation in the Anthropocene," *Science* 345, no. 6195 (2014): 401–6.

2. TURN YOUR DOG'S PAW UPSIDE DOWN

1. Daniele Silvestro et al., "The Role of Clade Competition in the Diversification of North American Canids," *Proceedings of the National Academy of Science* 112, no. 28 (2015): 8684–89. The evolution of cats and their entrance into North America are discussed in Lars Werdelin and Nobuyuki Yamaguchi, "Phylogeny and Evolution of Cats (Felidae)," *Biology and Conservation of Wild Felids*, ed. D. M. MacDonald and A. Loveridge (New York: Oxford University Press, 2010), 59–82. A good source for the evolution of canines is Xiaoming Wang and Richard H. Tedford, *Dogs: Their Fossil Relatives and Evolutionary History* (New York: Columbia University Press, 2008). I also used José R. Castelló, *Canids of the World: Wolves, Wild Dogs, Foxes, Jackals, Coyotes, and Their Relatives* (Princeton, NJ: Princeton University Press, 2018).
2. A number of sources on the internet continue to say that the native North American red fox is the introduced European red fox. This is not true. Nor is today's North American red fox genetically influenced much by the introduced European red fox, according to studies like Mark J. Stratham et al., "The Origin of Recently Established Red Fox Populations in the United States: Translocations or Natural Range Expansions?," *Journal of Mammalogy* 93, no. 1 (2012): 52–65.

3. The Pleistocene extinctions are the source of much lively discussion, and I have written about them before in my *When the Land Was Young: Reflections on American Archeology* (New York: Open Road Integrated Media, 2022). My current favorite analysis is Ross D. E. McPhee, *End of the Megafauna: The Fate of the World's Hugest, Fiercest, and Strangest Animals* (New York: Norton, 2019). Regarding dates, McPhee notes, "Credible last appearance (terminal) dates for North American species lost during the Pleistocene-Holocene transition mostly fall within a range of 10,500 to 11,000 radiocarbon years BP. Using a specific set of assumptions, this interval calibrates to a somewhat broader span in calendar years, 12,000 to 13,000 calendar years ago" (191).

3. COYOTES ARE THE ORIGINAL AIKIDO

1. Dan Flores, *Coyote America* (New York: Basic Books, 2016), 25. Examples of Coyote tales are also from Carobeth Laird, *Mirror and Pattern* (Banning, CA: Malki Museum Press, 1984); and Carobeth Laird, *The Chemehuevis* (Banning, CA: Malki Museum Press, 1976).
2. Jonathan G. Way and William S. Lynn, "Northeastern Coyote/Coywolf Taxonomy and Admixture: A Meta-Analysis," *Canid Biology and Conservation* 19, no. 1 (2016): 1–7.
3. A number of sources discuss this behavior of coyotes. You can see for yourself by watching Chicago researcher Stanley Gehrt in Gehrt, "Urban Coyotes—First Friday Watch Party," streamed live on December 3, 2021, YouTube video, 30:00, https://www.youtube.com/watch?v=sCZhnRbQlaY.
4. Monisha Ravisetti, "Scientists Now Know Why Coyotes Killed a Human in 2009," *CNET*, December 18, 2022, https://www.cnet.com/science/biology/scientists-now-know-why-coyotes-unexpectedly-killed-a-human-in-2009/.
5. Mark Elbroch and Kurt Rinehart, *Behavior of North American Mammals* (Boston: Houghton Mifflin, 2011), 76. The natural history of animals in *What Walks This Way* also comes from other sources, including my own experiences.

4. LET'S JUST ENJOY FOXES

1. Mark Elbroch and Kurt Rinehart, *Behavior of North American Mammals* (Boston: Houghton Mifflin, 2011), 100.
2. Lorraine Boissoneault, "Foxes and Coyotes Are Natural Enemies. Or Are They?" *Smithsonian*, March 8, 2018, https://www.smithsonianmag.com/science-nature/foxes-and-coyotes-are-natural-enemies-or-are-they-180968424/. Also see Marcus A. Mueller et al., "Co-existence of Coyotes (*Canis latrans*) and Red Foxes (*Vulpes vulpes*) in an Urban Landscape," *PLOS One*, January 24, 2018, https://doi.org/10.1371/journal.pone.019097.

5. IT ALMOST SEEMS WOLVES SHOULD BE ALLOWED TO VOTE

1. Alexander Kock et al., "Earth System Impacts of the European Arrival and Great Dying in the Americas After 1492," *Quaternary Science Reviews* 207, no. 1 (March 2019): 13–36.
2. Brodie Farquhar, "Wolf Reintroduction Changes Ecosystems in Yellowstone," Yellowstone National Park Trips, June 22, 2023, https://www.yellowstonepark.com/things-to-do/wildlife/wolf-reintroduction-changes-ecosystem/. Also see Rick McIntyre, *The Rise of Wolf Eight* (Vancouver: Graystone, 2019); and Nate Blakeslee, *American Wolf* (New York: Crown, 2017).
3. Rick McIntyre, *The Reign of Wolf 21* (Vancouver: Graystone, 2020), 21.
4. Gordon Haber and Marybeth Holleman, *Among Wolves: Gordon Haber's Insights Into Alaska's Most Misunderstood Animal* (Anchorage: University of Alaska Press, 2013), 257.

6. WE'LL LIFT THAT LION RIGHT OFF THAT LIMB

1. Robert McCurdy, *Life of the Greatest Guide* (Phoenix: Blue River Graphics, 1979), 39–41.
2. McCurdy, *Life of the Greatest Guide*, 36.
3. McCurdy, *Life of the Greatest Guide*, 235.
4. Thomas L. Servass et al., "North American Model of Wildlife Conservation Empowerment and Exclusivity Hinder Advances in Wildlife Conservation," *Canadian Wildlife Biology and Management* 7, no. 2 (2018): 101–18. Also see M. Nils Petersen and Michael Paul Nelson, "Why the North American Model of Wildlife Conservation Is Problematic for Modern Wildlife Management," *Human Dimensions of Wildlife* 22, no. 1 (2017): 43–54.
5. Harley Shaw, personal correspondence.

7. BOBCAT AND LYNX

1. Paul Rezendes, *Tracking and the Art of Seeing* (New York: Harper Collins, 1999), 15.
2. Mark Elbroch, *Mammal Tracks and Sign: A Guide to North American Species* (Mechanicsburg, PA: Stackpole, 2003), 3.
3. Quoted in Elbroch, *Mammal Tracks and Sign*, 3. Also see Louis Liebenberg, *The Art of Tracking: The Origin of Science* (Capetown: New Africa, 2012).

8. A BLACK-AND-WHITE AESTHETIC

1. Quoted in Lynne Warren, "Skunks: Notorious or Not?," National Wildlife Federation, March 27, 2017, https://www.nwf.org/Magazines/National-Wildlife/2017/April-May/Animals/Skunks.

9. THE ENTANGLED WORLD OF HUMANS AND RACCOONS

1. CBC News, "Urban Diet Leading to Fat Hypoglycemic Raccoons," July 3, 2018, https://www.cbc.ca/news/canada/sudbury/fat-urban-raccoons-diet-obesity-1.4731574.
2. Jill Riddell, "The City Raccoon and the Country Raccoon," *Chicago Wilderness*, Summer 2002, 29.
3. CBS News, "Are We Making Raccoons Smarter?," November 29, 2020, https://www.cbsnews.com/news/are-we-making-raccoons-smarter/.
4. Riddell, "The City Raccoon and the Country Raccoon," 29.
5. Liz Langley, "Raccoons Pass Famous Intelligence Test—By Upending It," *National Geographic*, October 21, 2017, https://www.nationalgeographic.com/animals/article/animals-intelligence-raccoons-birds-aesops. Also see Nat Geo Wild, "See Raccoons Pass Famous Intelligence Test," YouTube, https://www.youtube.com/watch?v=DZeVHcm4is4.
6. Veronica Pacini-Ketchabaw and Fikile Nxumalo, "Unruly Raccoons and Troubled Educators: Nature/Culture Divides in a Childcare Centre," *Environmental Humanities* 7 (2015): 152.
7. Ilima Loomis, "Raccoon Latrines Could Have a Hidden Impact on Ecosystems," *Science*, January 3, 2018.

10. THREE BEARS

1. Michael Shea, "The Return of the Grizzly, and Bear Hunting, in the West," *Outdoor Life*, October 22, 2021, https://www.outdoorlife.com/hunting-grizzly-bears-necessary-conservation/.
2. Kyle A. Artelle, "Is Wildlife Conservation Policy Based in Science?," *American Scientist* 107, no. 1 (January–February 2019): 38–50.
3. Kyle A. Artelle et al., "Hallmarks of Science Missing from North American Wildlife Management," *Science Advances* 4, no. 3 (2018).
4. David J. Mattson, "Grizzly Bears for the Southwest," The Grizzly Bear Project, https://www.allgrizzly.org/.
5. Emma Marris, *Wild Souls* (New York: Bloomsbury, 2021), 7.

11. THE SUDDEN BEATING OF BRAINS

1. Annie Dillard, *Teaching a Stone to Talk* (New York: Harper Perennial, 1982), 67.
2. Dillard, *Teaching a Stone to Talk*, 69–70.

12. YOU THINK YOU KNOW WHAT THE IUCN IS GOING TO SAY

1. Quoted in Mark Olalde, "Deadly Virus in Rabbits Threatens to Upend Some Western Ecosystems," *Palm Springs Desert Sun*, April 16, 2021.

2. Andrew T. Smith, "Conservation Status of American Pikas (*Ochotona princeps*), *Journal of Mammalogy* 101, no. 6 (2020): 1466–88.

13. THE LITTLE GUYS

1. Mark Elbroch, *Mammal Tracks and Sign: A Guide to North American Species* (Mechanicsburg, PA: Stackpole, 2003), 340.
2. Clark E. Adams, *Urban Wildlife Management*, 3rd ed. (Boca Raton, FL: CRC, 2016), 178.

14. HOOVES

1. Olivia Laing, "The Wild Beauty of Georgia O'Keefe," *Guardian*, July 1, 2016, https://www.theguardian.com/artanddesign/2016/jul/01/georgia-okeeffe-tate-modern-exhibition-wild-beauty.
2. Colorado Parks and Wildlife, "The Story of Colorado's Mule Deer," September 2017, https://cpw.state.co.us/documents/muledeer/coloradosmuledeerstory.pdf.
3. Bruce Finley, "Colorado Embarks on Experimental 'Predator Control' Killing of More Lions and Bear to Try and Save Dwindling Deer," *Denver Post*, December 14, 2016.
4. Kristen A. Schmitt, "Winter Weather Causes Colorado Wildlife to Struggle," March 08, 2023, https://www.gohunt.com/content/news/winter-weather-causes-colorado-wildlife-to-struggle.
5. David Tenenbaum, "Deer Account for Almost Half of Long-Term Forest Changes, Study Finds," University of Wisconsin–Madison, January 2, 2015, https://news.wisc.edu/deer-account-for-almost-half-of-long-term-forest-change-study-finds/. Also see Brian Payton, "It's Time We Talked About Our Bambi Problem," *Mother Jones*, February 11, 2023, https://www.motherjones.com/politics/2023/02/bambi-deer-overpopulation-environmental-damage/.
6. David Milbank, "I Bought a Gun and I Intend to Use It," *Washington Post*, February 17, 2023.
7. Aldo Leopold, *A Sand County Almanac* (New York: Random House, 1966), 139–40.
8. Paul Liotta, "Vasectomy Strategy Cuts Staten Island Deer Population by 30 Percent; Encouraging Data on Multiple Fronts," SIlive.com, December 28, 2022, https://www.silive.com/news/2022/12/vasectomy-strategy-cuts-staten-islands-deer-population-by-30-encouraging-data-on-multiple-fronts.html.

15. SKULLS AND BONES

1. Mark Elbroch, *Animal Skulls: A Guide to North American Species* (Mechanicsburg, PA: Stackpole, 2006), 11.
2. Roxana Robinson, *Georgia O'Keefe: A Life* (Lebanon, NH: University Press of New England, 1999), 365.

16. A NEW VISION OF WILDLIFE MANAGEMENT

1. Wildlife for All, "Wildlife for All National Campaign Launched to Transform Wildlife Conservation," October 1, 2021, https://wildlifeforall.us/wildlife-for-all-national-campaign-launched-to-transform-wildlife-conservation/.

17. THE EVALUATION

1. Jonah Evans, "Determining Observer Reliability in Counts of River Otter Tracks," *Journal of Wildlife Management* 73, no. 3 (April 2009): 426–32.

Bibliography

Adams, Clark E. *Urban Wildlife Management*, 3rd ed. Boca Raton, FL: CRC Press, 2016.

Artelle, Kyle A. "Is Wildlife Conservation Policy Based in Science?" *American Scientist* 107, no. 1 (January–February 2019): 38–50.

Artelle, Kyle A., John D. Reynolds, Adrian Treves, Jessica C. Walsh, Paul C. Paquet, and Chris T. Darimont. "Hallmarks of Science Missing from North American Wildlife Management." *Science Advances* 4, no. 3 (2018).

Bar-On, Yinon M. "The Biomass Distribution on Earth." *Proceedings of the National Academy of Science* 115, no. 25 (2018): 6506–11.

Blakeslee, Nick. *American Wolf.* New York: Crown, 2017.

Boissoneault, Lorraine. "Foxes and Coyotes Are Natural Enemies. Or Are They?" *Smithsonian*, March 8, 2018.

Brookshire, Bethany. "Coyotes Came to New York City but Not for Our Pizza." *New York Times*, October 6, 2022.

Castelló, José R. *Canids of the World: Wolves, Wild Dogs, Foxes, Jackals, Coyotes, and Their Relatives.* Princeton, NJ: Princeton University Press, 2018.

CBC News. "Urban Diet Leading to Fat Hypoglycemic Raccoons." July 3, 2018. https://www.cbc.ca/news/canada/sudbury/fat-urban-raccoons-diet-obesity-1.4731574.

CBS News. "Are We Making Raccoons Smarter?" November 29, 2020. https://www.cbsnews.com/news/are-we-making-raccoons-smarter/.

Colorado Parks and Wildlife. "The Story of Colorado's Mule Deer." September 2017. https://cpw.state.co.us/documents/muledeer/coloradosmuledeerstory.pdf.

Dillard, Annie. *Teaching a Stone to Talk.* New York: Harper Perennial, 1982.

Dirzo, Rodolfo, Hillary S. Young, Mauro Galetti, Gerardo Ceballos, Nick J. B. Isaac, and Ben Collen. "Defaunation in the Anthropocene." *Science* 345, no. 6195 (2014): 401–6.

Dungan, Ron. "After the Devastation, Nature Fuels Recovery, Season After Season." *Republic*, June 22, 2012.

Elbroch, Mark. *Animal Skulls: A Guide to North American Species.* Mechanicsburg, PA: Stackpole, 2006.

——. *Mammal Tracks and Sign: A Guide to North American Species.* Mechanicsburg, PA: Stackpole, 2003.

Elbroch, Mark, with contributions by Casey McFarland. *Mammal Tracks and Sign: A Guide to North American Species.* Mechanicsburg, PA: Stackpole, 2019.

Elbroch, Mark, and Kurt Rinehart. *Behavior of North American Mammals.* Boston: Houghton Mifflin, 2011.

Evans, Jonah. "Determining Observer Reliability in Counts of River Otter Tracks." *Journal of Wildlife Management* 73, no. 3 (April 2009): 426–32.

Farquhar, Brodie. "Wolf Reintroduction Changes Ecosystems in Yellowstone." Yellowstone National Park Trips, June 22, 2023. https://www.yellowstonepark.com/things-to-do/wildlife/wolf-reintroduction-changes-ecosystem/.

Fascione, Nina, Aimee Delach, and Martin Smith, eds. *People and Predators, from Conflict to Coexistence.* Washington, DC: Island Press, 2004.

Finley, Bruce. "Colorado Embarks on Experimental 'Predator Control' Killing of More Lions and Bear to Try and Save Dwindling Deer." *Denver Post,* December 14, 2016.

Flores, Dan. *Coyote America.* Basic Books, 2016.

Gehrt, Stanley, Seth P. D. Riley, and Brian L Cypher, eds. *Urban Carnivores: Ecology, Conflict, and Conservation.* Baltimore: Johns Hopkins University Press, 2010.

Grayson, Donald K. *Giant Sloths and Sabertooth Cats: Extinct Animals and the Archaeology of the Ice Age Great Basin.* Salt Lake City: University of Utah Press, 2016.

Haber, Gordon, and Marybeth Holleman. *Among Wolves: Gordon Haber's Insights Into Alaska's Most Misunderstood Animal.* Anchorage: University of Alaska Press, 2013.

Hansen, Kevin. *Bobcat: Master of Survival.* New York: Oxford University Press, 2007.

Kock, Alexander, Chris Brierley, Mark M. Maslin, and Simon L. Lewis. "Earth System Impacts of the European Arrival and Great Dying in the Americas After 1492. *Quaternary Science Reviews* 207, no. 1 (March 2019): 13–36.

Kwasny, Melissa. *Putting on the Dog: Animal Origins of What We Wear.* San Antonio, TX: Trinity University Press, 2022.

Laing, Olivia. "The Wild Beauty of Georgia O'Keefe." *Guardian,* July 1, 2016.

Laird, Carobeth. *The Chemehuevis.* Banning, CA: Malki Museum Press, 1976.

——. *Mirror and Pattern.* Banning, CA: Malki Museum Press, 1984.

Langley, Liz. "Raccoons Pass Famous Intelligence Test—By Upending It." *National Geographic,* October 21, 2017. https://www.nationalgeographic.com/animals/article/animals-intelligence-raccoons-birds-aesops.

Leathlobhair, Máire Ní, et al. "The Evolutionary History of Dogs in America." *Science* 361 (2018): 81–85.

Leopold, Aldo. *A Sand County Almanac.* New York: Random House, 1966.

Liebenberg, Louis. *The Art of Tracking: The Origin of Science.* Capetown: New Africa, 2012.

Liotta, Paul. "Vasectomy Strategy Cuts Staten Island Deer Population by 30 Percent; Encouraging Data on Multiple Fronts." SIlive.com, December 28, 2022.

Loomis, Ilima. "Raccoon Latrines Could Have a Hidden Impact on Ecosystems." *Science,* January 3, 2018.

Marris, Emma. *Wild Souls.* New York: Bloomsbury, 2021.

Mattson, David J. "Grizzly Bears for the Southwest." The Grizzly Bear Project. https://www.allgrizzly.org/.

McCurdy, Robert. *Life of the Greatest Guide*. Phoenix: Blue River Graphics, 1979.

McIntyre, Rick. *The Reign of Wolf 21*. Vancouver: Graystone, 2020.

——. *The Rise of Wolf Eight*. Vancouver: Graystone, 2019.

McPhee, Ross D. E. *End of the Megafauna: The Fate of the World's Hugest, Fiercest, and Strangest Animals*. New York: Norton, 2019.

Milbank, David. "I Bought a Gun and I Intend to Use It." *Washington Post*, February 17, 2023.

Mueller, Marcus A., David Drake, and Maximilian L. Allen. "Co-existence of Coyotes (*Canis latrans*) and Red Foxes (*Vulpes vulpes*) in an Urban Landscape." *PLOS One*, January 24, 2018.

Olalde, Mark. "Deadly Virus in Rabbits Threatens to Upend Some Western Ecosystems." *Palm Springs Desert Sun*, April 16, 2021.

Pacini-Ketchabaw, Veronica, and Fikile Nxumalo. "Unruly Raccoons and Troubled Educators: Nature/Culture Divides in a Childcare Centre." *Environmental Humanities* 7 (2015): 151–68.

Payton, Brian. "It's Time We Talked About Our Bambi Problem." *Mother Jones*, February 11, 2023. https://www.motherjones.com/politics/2023/02/bambi-deer-overpopulation-environmental-damage/.

Petersen, Nils, and Michael Paul Nelson. "Why the North American Model of Wildlife Conservation Is Problematic for Modern Wildlife Management." *Human Dimensions of Wildlife* 22, no. 1 (2017): 43–54.

Prothero, Donald R. *The Princeton Field Guide to Prehistoric Mammals*. Princeton, NJ: Princeton University Press, 2017.

Rezendes, Paul. *Tracking and the Art of Seeing*. New York: Harper Collins, 1999.

Riddell, Jill. "The City Raccoon and the Country Raccoon." *Chicago Wilderness*, Summer 2002, 28–29.

Robinson, Roxanna. *Georgia O'Keefe: A Life*. Lebanon, NH: University Press of New England, 1999.

Russell, Sharman Apt. *Diary of a Citizen Scientist: Chasing Tiger Beetles and Other New Ways of Engaging with the World*. New York: Open Road Integrated Media, 2022.

——. *Kill the Cowboy: A Battle of Mythology in the New West*. Boston: Addison-Wesley, 1993.

——. *When the Land Was Young: Reflections on American Archeology*. New York: Open Road Integrated Media, 2022.

Schmitt, Kristen A. "Winter Weather Causes Colorado Wildlife to Struggle." GoHunt, March 28, 2023. https://www.gohunt.com/content/news/winter-weather-causes-colorado-wildlife-to-struggle.

Servass, Thomas L., Robert P. Brooks, and Jeremy T. Bruskotter. "North American Model of Wildlife Conservation: Empowerment and Exclusivity Hinder Advances in Wildlife Conservation." *Canadian Wildlife Biology and Management* 7, no. 2 (2018): 101–18.

Shea, Michael. "The Return of the Grizzly, and Bear Hunting, in the West." *Outdoor Life*, October 22, 2021. https://www.outdoorlife.com/hunting-grizzly-bears-necessary-conservation/.

Silvestro, Daniele, Alexandre Antonelli, Nicolas Salamin, and Tiago B. Quental. "The Role of Clade Competition in the Diversification of North American Canids." *Proceedings of the National Academy of Science* 112, no. 28 (2015): 8684–89.

Smith, Andrew T. "Conservation Status of American Pikas (*Ochotona princeps*)." *Journal of Mammalogy* 101, no. 6 (2020): 1466–88.

Stratham, Mark J., Benjamin N. Sacks, Keith B. Aubry, John D. Perrine, and Samantha M. Wisely. "The Origin of Recently Established Red Fox Populations in the United States: Translocations or Natural Range Expansions?" *Journal of Mammalogy* 93, no. 1 (2012): 52–65.

Tenenbaum, David. "Deer Account for Almost Half of Long-Term Forest Changes, Study Finds." University of Wisconsin–Madison, January 2, 2015. https://news.wisc.edu/deer-account-for-almost-half-of-long-term-forest-change-study-finds/.

Wang, Xiaoming, and Richard H. Tedford. *Dogs: Their Fossil Relatives and Evolutionary History*. New York: Columbia University Press, 2008.

Warren, Lynne. "Skunks: Notorious—or Not?" National Wildlife Federation, March 27, 2017. https://www.nwf.org/Magazines/National-Wildlife/2017/April-May/Animals/Skunks.

Way, Jonathan G., and William S. Lynn. "Northeastern Coyote/Coywolf Taxonomy and Admixture: A Meta-Analysis." *Canid Biology and Conservation* 19, no. 1 (2016): 1–7.

Werdelin, Lars, and Nobuyuki Yamaguchi. "Phylogeny and Evolution of Cats (Felidae)." In *Biology and Conservation of Wild Felids*, ed. D. M. MacDonald and A. Loveridge, 59–82. New York: Oxford University Press. 2010.

Wildlife for All. "Wildlife for All National Campaign Launched to Transform Wildlife Conservation." October 1, 2021. https://wildlifeforall.us/wildlife-for-all-national-campaign-launched-to-transform-wildlife-conservation/.

World Wildlife Fund. *Living Planet Report 2022*. https://www.worldwildlife.org/pages/living-planet-report-2022.

Zeveloff, Samuel I. *Raccoons: A Natural History*. Washington, DC: Smithsonian Institution Press, 2002.

Index

Page numbers in *italics* refer to images.

Printed in the USA
CPSIA information can be obtained
at www.ICGtesting.com
JSHW022003160824
68283JS00001B/14